How Math Explains the World

Smithsonian Books

Collins
An Imprint of HarperCollins*Publishers*

How Math Explains the World

A Guide

to the

Power of Numbers,

from

Car Repair to

Modern Physics

James D. Stein

HarperCollins books may be purchased for educational, business, or sales promotional use.
For information, please write: Special Markets Department, HarperCollins Publishers, 10 East
53rd Street, New York, NY 10022.

FIRST EDITION

Designed by Sunil Manchikanti

Library of Congress Cataloging-in-Publication Data

Stein, James D.
  How math explains the world : a guide to the power of numbers, from car repair to modern
physics / Jim Stein.
    p. cm.
  ISBN: 978-0-06-124176-5
    1. Mathematics—Philosophy.   2. Mathematics—Psychological aspects.
3. Mathematics—Sociological aspects.   4. Number concept.   5. Mathematical
ability. I. Title.
  QA8.4.S72 2008
  510—dc22

                                              2007041784

11 WBC/RRD 10 9 8 7 6 5 4

To my lovely wife, Linda,
for whom no dedication does justice.

Contents

Preface

The November Statement

My first glimpse into mathematics, as opposed to arithmetic, came on a Saturday afternoon in late fall when I was about seven years old. I wanted to go out and toss a football around with my father. My father, however, had other ideas.

For as long as I can remember, my father always kept a meticulous record of his monthly expenses on a large yellow sheet that, in retrospect, was a precursor of an Excel spreadsheet. One yellow sheet sufficed for each month; at the top, my father wrote the month and year, and the rest of the sheet was devoted to income and expenses. On this particular fall day, the sheet had failed to balance by 36 cents, and my father wanted to find the discrepancy.

I asked him how long it would take, and he said he didn't think it would take too long, because errors that were divisible by 9 were usually the result of writing numbers down in the wrong order; writing 84 instead of 48; $84 - 48 = 36$. He said this always happened; whenever you wrote down a two-digit number, reversed the digits, and subtracted one from the other, the result was always divisible by 9.[1]

Seeing as I wasn't going to be able to toss a football around for a while, I got a piece of paper and started checking my father's statement. Every number I tried worked; $72 - 27 = 45$, which was divisible by 9. After a while, my father found the error; or at least decided that maybe he should play football with me. But the idea that there were patterns in numbers took root in my mind; it was the first time that I realized there was more to arithmetic than the addition and multiplication tables.

Over the years, I have learned about mathematics and related subjects from four sources. In addition to my father, who was still attending Sunday-morning mathematics lectures when he was in his seventies, I was fortunate to have some excellent teachers in high school, college, and graduate school. When the Russians launched Sputnik in 1957, schools scrambled desperately to prepare students for careers in science and engineering; the Advanced Placement courses took on added importance. I was in one of the first such courses, and took a wonderful course in calculus my senior year in high school from Dr. Henry Swain. One of my regrets is that I never got a chance to tell him that I had, to some extent, followed in his footsteps.

In college I took several courses from Professor George Seligman, and I was delighted to have the opportunity to communicate with him as I was writing this book. However, the greatest stroke of good fortune in my career was to have Professor William Bade as my thesis adviser. He was not only a wonderful teacher, but an inspired and extremely tolerant mentor, as I was not the most dedicated of graduate students (for which I blame an addiction to duplicate bridge). The most memorable day of my graduate career was not the day I finished my thesis, but the day Bill received a very interesting and relevant paper.[2] We met at two and started going over the paper, broke for dinner around 6:30, and finished, somewhat bleary-eyed, around midnight. The paper itself was a breakthrough in the field, but the experience of going through it, discussing the mathematics and speculating on how I might use it to develop a thesis, made me realize that this was something I wanted to do.

There are a number of authors whose books had a profound effect on me. There are too many to list, but the most memorable books were George Gamow's *One, Two, Three . . . Infinity,* Carl Sagan's *Cosmos,* James Burke's *Connections,* John Casti's *Paradigms Lost,* and Brian Greene's *The Elegant Universe* and *The Fabric of the Cosmos.* Only two of these books were published during the same decade, which attests to a long-standing tradition of excellence in science writing. I'd be happy if this book was mentioned in the same breath as any of the above.

I've had many colleagues over the years with whom I've discussed math and science, but two in particular stand out: Professors Robert Mena and Kent Merryfield at California State University, Long Beach. Both are excellent mathematicians and educators with a far greater knowledge and appreciation of the history of mathematics than I have, and writing this book was made considerably easier by their contributions.

There have been several individuals of varying technical backgrounds with whom I have had illuminating conversations. My understanding of some of the ideas in this book was definitely helped by conversations with Charles Brenner, Pete Clay, Richard Helfant, Carl Stone, and David Wilczynski, and I am grateful to all of them for helping me to think through some of the concepts and devising different ways of explaining them.

Finally, I'd like to thank my agent, Jodie Rhodes, without whose persistence this book may never have seen the light of day, and my editor, T. J. Kelleher, without whose suggestions both the structure and the presentations in this book would have been much less coherent—T.J. has the rare gift of improving a book on both the macro and the micro level. And, of course, my wife, Linda, who contributed absolutely nothing to the book, but contributed inestimably to all other aspects of my life.

NOTES

1. Any two-digit number can be written as $10T + U$, where T is the tens digit and U the units digit. Reversing the digits gives the number $10U + T$, and subtracting the second from the first yields $10T + U - (10U + T) = 9T - 9\ U = 9(T - U)$, which is clearly divisible by 9.
2. B. E. Johnson, "Continuity of Homomorphisms of Algebras of Operators," *Journal of the London Mathematical Society*, 1967: pp. 537–541. It was only four pages long, but reading research mathematics is not like reading the newspaper. Although it was not a technically difficult paper (no involved calculations, which can slow down the pace of reading to a crawl), it contained a number of incredibly ingenious ideas that neither Bill nor I had seen before. This paper essentially made my thesis, as I was able to adapt some of Johnson's ideas to the problem that I had been addressing.

Introduction

Not Just a Rock

We advance, both as individuals and as a species, by solving problems. As a rule of thumb, the reward for solving problems increases with the difficulty of the problem. Part of the appeal of solving a difficult problem is the intellectual challenge, but a reward that often accompanies the solutions to such problems is the potential to accomplish amazing feats. After Archimedes discovered the principle of the lever, he remarked that if he were given a lever and a place to stand, he could move Earth.[1] The sense of omnipotence displayed in this statement can also be found in the sense of omniscience of a similar observation made by the eighteenth-century French mathematician and physicist Pierre-Simon de Laplace. Laplace made major contributions to celestial mechanics, and stated that if he knew the position and velocity of everything at a given moment, he would be able to predict where everything would be at all times in the future.

"Given for one instant an intelligence which could comprehend all the forces by which nature is animated and the respective positions of the beings which compose it, if moreover this intelligence were vast enough to submit these data to analysis, it would embrace in the same formula both the movements of the largest bodies in the universe and those of the lightest atom; to it nothing would be uncertain, and the future as the past would be present to its eyes."[2]

Of course, these statements were rhetorical, but they were made to emphasize the far-reaching potential of the solution to the problem. A casual onlooker, seeing Archimedes use a lever to reposition a heavy rock, might have said, "OK, that's useful, but it's just a rock." Archimedes could have replied, "It's not just this rock—it's any object whatsoever, and I can tell you what length lever I need to move that object and how much effort I will have to exert in order to move the object to a desired position."

Sometimes we are so impressed with the more dazzling achievements of science and engineering that our inability to solve seemingly easy (or easier) problems appears puzzling. During the 1960s, one could occasionally hear the following comment: If they can put a man on the moon, how come they can't cure the common cold?

We are a little more scientifically sophisticated now, and most people are willing to cut science some slack on problems like this, recognizing that curing the common cold is a more difficult problem than it initially seems. The general feeling, though, is that we just haven't found a cure for the common cold—yet. It's obviously a difficult problem, but considering the potential payoff, it's no surprise that medical researchers are busily trying, and most of us would probably expect them to find a cure sooner or later. Sadly, for those suffering from runny noses and sore throats, there is a very real possibility that a cure for the common cold may never be found, not because we aren't clever enough to find it, but because it may not exist. One of the remarkable discoveries of the twentieth century is a common thread that runs through mathematics, the natural sciences, and the social sciences—there are things that we cannot know or do, and problems that are incapable of solution. We know, and have known for some time, that humans are neither omnipotent nor omniscient, but we have only recently discovered that omnipotence and omniscience may simply not exist.

When we think of the scientific developments of the twentieth century, we think of the giant strides that were made in practically every discipline, from astronomy through zoology. The structure of DNA. The theory of relativity. Plate tectonics. Genetic engineering. The expanding universe. All of these breakthroughs have contributed immeasurably to our knowledge of the physical universe, and some have already had a significant impact on our daily lives. This is the great appeal of science—it opens doors for us to learn fascinating things and, even better, to use what we learn to make our lives richer beyond imagining.

However, the twentieth century also witnessed three eye-opening results that demonstrated how there are limits—limits to what we can know and do in the physical universe, limits to what truths we can discover using mathematical logic, and limits to what we can achieve in implementing democracy. The most well-known of the three is Werner Heisenberg's uncertainty principle, discovered in 1927. The uncertainty principle shows that not even an individual possessed of omniscience could have supplied Laplace with the positions and velocities of all the objects in the universe, because the positions and velocities of those objects cannot be simultaneously determined. Kurt Gödel's incompleteness theorem, proved a decade later, reveals the inadequacy of logic to determine mathematical truth. Roughly fifteen years after Gödel established the incompleteness theorem, Kenneth Arrow showed that there is no method of tabulating votes that can satisfactorily translate the preferences of the individual voters into the preferences of the society to which those voters belong. The second half of the twentieth century witnessed a profusion of results in a number of areas, demonstrating how our ability to know and to do is limited, but these are unquestionably the Big Three.

There are a number of common elements to these three results. The first is that they are all mathematical results, whose validity has been established by mathematical proof.

It is certainly not surprising that Gödel's incompleteness theorem, which is obviously a result about mathematics, was established through mathematical argument. It is also not surprising that Heisenberg's uncertainty principle is the result of mathematics—we have been taught since grade school that mathematics is one of the most important tools of science, and physics is a discipline that relies heavily on mathematics. However, when we think of the social sciences, we do not usually think of mathematics. Nonetheless, Arrow's theorem is completely mathematical, in a sense even more so than Heisenberg's uncertainty principle, which is a mathematical result derived from hypotheses about the physical world.

Arrow's theorem is as "pure" as the "purest" of mathematics—it deals with functions, one of the most important mathematical concepts. Mathematicians study all types of functions, but the properties of the functions studied are sometimes dictated by specific situations. For instance, a surveyor would be interested in the properties of trigonometric functions, and might embark upon a study of those functions realizing that knowledge of their properties could help with problems in surveying. The properties of the functions discussed in Arrow's theorem are clearly motivated by the problem Arrow initially started to investigate—how to translate the preferences of individuals (as expressed by voting) into the results of an election.

The utility of mathematics is due in large measure to the wide variety of situations that are amenable to mathematical analysis. The following tale has been repeated time and time again—some mathematician does something that seems of technical interest only, it sits unexamined for years (except possibly by other mathematicians), and then somebody finds a totally unexpected practical use for it.

An instance of this situation that affects practically everyone in the civilized world almost every day would have greatly surprised G. H. Hardy, an eminent British mathematician who lived during the first half of the twentieth century. Hardy wrote a fascinating book *(A Mathematician's Apology)*, in which he described his passion for the aesthetics of mathematics. Hardy felt that he had spent his life in the search for beauty in the patterns of numbers, and that he should be regarded in the same fashion as a painter or a poet, who spends his or her life in an attempt to create beauty. As Hardy put it, "a mathematician, like a painter or a poet, is a maker of patterns. If his patterns are more permanent than theirs, it is because they are made with ideas."[3]

Hardy made great contributions to the theory of numbers, but viewed his work and that of his colleagues as mathematical aesthetics—possessing beauty for those capable of appreciating it, but having no practical value. "I have never done anything 'useful'. No discovery of mine has made, or is likely to make, directly or indirectly, for good or ill, the least difference to the amenity of the world,"[4] he declared, and undoubtedly felt the same way about his coworkers in number theory. Hardy did not foresee that within fifty years of his death, the world would rely heavily on a phenomenon that he spent a good portion of his career investigating.

Prime numbers are whole numbers that have no whole number divisors other than 1 and the number itself; 3 and 5 are primes, but 4 is not because it is divisible by 2. As one looks at larger and larger numbers, the primes become relatively more infrequent; there are 25 primes between 1 and 100, but only 16 between 1,000 and 1,100, and only 9 between 7,000 and 7,100. Because prime numbers become increasingly rare, it becomes extremely difficult to factor very large numbers that are the product of two primes, in the sense that it takes a lot of time to find the two primes that are the factors (a recent experiment took over nine months with a large network of computers). We rely on this fact every day, when we type in a password or take money from an ATM, because this difficulty in factoring large numbers that are the product of two primes is the cornerstone of many of today's computerized security systems.

Like number theory, each of the Big Three has had a profound, although somewhat delayed, impact. It took a while, but the uncertainty principle, and the science of quantum mechanics of which it is a part, has brought us most of the microelectronic revolution—computers, lasers, magnetic resonance imagers, the whole nine yards. The importance of Gödel's theorem was not initially appreciated by many in the mathematical community, but that result has since spawned not only branches of mathematics but also branches of philosophy, extending both the variety of the things we know, the things we don't, and the criteria by which we evaluate whether we know or can know. Arrow did not receive a Nobel Prize until twenty years after his theorem was first published, but this result has significantly expanded both the range of topics and the methods of studying those topics in the social sciences, as well as having practical applications to such problems as the determination of costs in network routing problems (how to transmit a message from Baltimore to Beijing as cheaply as possible).

Finally, a surprising common element uniting these three results is that they are—well, surprising (although mathematicians prefer the word *counterintuitive*, which sounds much more impressive than *surprising*). Each of these three results was an intellectual bombshell, exploding preconceptions held by many of the leading experts in their respective fields. Heisenberg's uncertainty principle would have astounded Laplace and the many other physicists who shared Laplace's deterministic vision of the universe. At the same mathematics conference that David Hilbert, the leading mathematician of the day, was describing to a rapt audience his vision of how mathematical truth might some day be automatically ascertained, in a back room far from the limelight Gödel was showing that there were some truths whose validity could never be proven. Social scientists had searched for the ideal method of voting even before the success of the American and French Revolutions, yet before he even finished graduate school, Arrow was able to show that this was an impossible goal.

The Difficult We Do Today, but the Impossible Takes Forever

There is a fairly simple problem that can be used to illustrate that something is impossible. Suppose that you have an ordinary eight-by-eight chessboard and a supply of tiles. Each tile is a rectangle whose length is twice the length of a single square of the chessboard, and whose width is the length of one square of the chessboard, so that each tile covers exactly two adjacent squares of the chessboard.

It is easy to cover the chessboard exactly with 32 tiles, so that all squares are covered and no tile extends beyond the boundary of the chessboard. Since each row can be covered by laying four tiles end to end, do that for each of the eight rows. Now, suppose that you remove the two squares at the ends of a long diagonal from the chessboard; these might be the square at the left end of the back row and the square at the right end of the front row. This leaves a board that has only 62 squares remaining. Can you cover this board exactly with 31 tiles, so that every square is covered?

As you might suspect from the lead-in to this section, or from some experimentation, this cannot be done; there is a simple, and elegant, reason for this. Imagine that the chessboard is colored in the usual way, with alternating black and red squares. Each tile covers precisely 1 black square and 1 red square, so the 31 tiles will cover 31 black squares and 31 red squares. If you look at a chessboard, the square at the left end of the back row and the square at the right end of the front row have the same color (we'll assume they are both black), so removing them leaves a board with 32 red squares and 30 black squares—which the 31 tiles cannot cover. It's a simple matter of counting; the clever part is seeing what to count.

One of the reasons for the power of both science and mathematics is that once a productive line of reasoning is established, there is a rush to extend the range of problems to which the line of reasoning applies. The above problem might be classed as a "hidden pattern"—it is obvious that each tile covers two squares, but without the coloring pattern normally associated with chessboards, it is not an easy problem to solve. Discovering the hidden pattern is often the key to mathematical and scientific discoveries.

When There Is No Music out There

We are all familiar with the concept of writer's block: the inability of a writer to come up with a good idea. The same thing can happen to mathematicians and scientists, but there is another type of block that exists for the mathematician or scientist for which there is no analogy from the arts. A mathematician or scientist may work on a problem that has no answer. A composer might be able to come to grips with the idea that, at the moment, he is incapable of composing music, but he would never accept the idea that there simply is no music out there to compose. Mathematicians and scientists are keenly aware that nature may frustrate all their efforts. Sometimes, there is no music out there.

Physics is currently embarked on a quest started by Albert Einstein, who spent perhaps the last half of his life in search of a unified field theory, which physicists now call a TOE, for theory of everything. Not all great physicists search for a TOE—Richard Feynman once remarked that, "If it turns out there is a simple ultimate law that explains everything, so be it. . . . If it turns out it's like an onion with millions of layers, and we're sick and tired of looking at layers, then that's the way it is."[5] Feynman may not have been looking for a TOE, but Einstein was, and many top physicists are.

Nevertheless, Einstein was almost certainly aware that there may be no TOE—simple and elegant, complicated and messy, or anything in between. During the latter portion of their careers, both Einstein and Gödel were at the Institute for Advanced Study in Princeton, New Jersey. Gödel, reclusive and paranoid, would talk only to Einstein. Given Gödel's proof, that some things are unknowable, it is a reasonable conjecture that they discussed the possibility that there was no unified field theory to be discovered, and that Einstein was chasing a wild goose. However, Einstein could afford to spend his creative efforts chasing a tantalizing wild goose—for he had made his reputation.

It may seem surprising that even those who work in mathematics and science without the credentials of an Einstein do not live in fear of working on a problem that turns out to be unsolvable. Such problems have occurred with some frequency throughout history—and quite often, even though the wild goose escapes, the result has not been failure, but the discovery of something new that is usually interesting and sometimes immensely practical. The stories of those "failures," and the surprising developments that occurred because of them, form the subject matter of this book.

Of Bank Robbers, Mathematicians, and Scientists

When asked why he robbed banks, Willie Sutton replied, "Because that's where the money is." Every mathematician or scientist dreams of making a remarkable discovery—not because that's where the money is (although fame and fortune do occasionally await those who make such discoveries), but because that's where the fascination is: to be the first to observe, or create, or understand something truly wonderful.

Even if that's where the wild geese lurk, we have a desperate need to solve some critical problems—and a strong desire to solve some intriguing ones—and the only way we are going to do that is to train highly competent people, and some brilliant ones, to attack those problems with the tools of mathematics and science. For centuries, we sought a philosopher's stone whose touch would transmute base metals into gold. We failed, but the desire to find the philosopher's stone led to the atomic theory and an understanding of chemistry, which allows us to reshape the material we find in the world to new and better uses. Is that not a much more desirable result for mankind than transmuting base metals into gold?

At the very least, learning what we cannot know and cannot do prevents us from needlessly expending resources in a futile quest—only Harry Potter would bother to search today for the philosopher's stone. We have no way of knowing—yet—if the quest for a TOE is the philosopher's stone search of today. However, if history is any guide, we will discover again that failing to find a wild goose might still lead us to a golden egg.

The Agent, the Editor, and Stephen Hawking's Publisher

In the introduction to his best seller *A Brief History of Time,* Stephen Hawking mentions that his publisher stated that for each equation he included, the readership would drop by 50 percent. Nonetheless, Hawking had sufficient confidence in his readership that he was willing to include Einstein's classic equation $E = mc^2$.

I'd like to think that the readers of this book are made of sterner stuff. After all, it's a book about mathematics, and equations represent not only great truths, such as the one in Einstein's equation, but the connecting threads that lead to those truths. In addition to Hawking's publisher, I have received input from my editor, who feels that mathematics is absolutely necessary in a book about mathematics, and my agent, who is happy to read about mathematics, but is decidedly unenthusiastic about reading mathematics.

There is clearly a fine line here, and so I have tried to write the book to allow those who want to skip a section in which mathematics is done to do so without losing the gist of what is being said. Those brave souls who want to follow the mathematics can do so with only a high-school mathematics background (no calculus). However, readers interested in pursuing the subject in greater depth can find references in the Notes (and occasionally a greater depth of treatment). In many instances, there is accessible material on the Web, and for most people it is easier to type in a URL than it is to chase something down in the library (especially since the neighborhood library is usually lacking in books outlining the mathematics of Galois theory or quantum mechanics). As a result, there are many references to Web sites in the appendix—but Web sites do disappear, and I hope the reader will forgive the occasional reference to such a site.

I hope that Hawking's publisher is wrong. If he is right and the population of the world is 6 billion, the thirty-third equation will reduce the potential readership for this book to less than a single reader.

NOTES

1. "Give me but one firm spot on which to stand, and I will move the Earth." *The Oxford Dictionary of Quotations*, 2nd ed. (London: Oxford University Press, 1953), p. 14.
2. Pierre-Simon de Laplace, *Theorie Analytique de Probabilites: Introduction*, v. VII, *Oeuvres (1812–1820)*.
3. G. H. Hardy, *A Mathematician's Apology*, public-domain version available at http://www.math.ualberta.ca/~mss/books/A%20Mathematician%27s%20Apology .pdf. This quote is from Section 10.
4. Ibid., Section 29.
5. *No Ordinary Genius: The Illustrated Richard Feynman*, ed. Christopher Sykes (New York: Norton, 1995).

How Math Explains the World

Prologue
Why Your Car Never Seems to Be Ready When They Promised

The $1 Million Questions

Every year a small collection of distinguished scientists, economists, literary giants, and humanitarians gather in Stockholm for the award of the prestigious—and lucrative—Nobel Prizes, and nary a mathematician is to be found among them. The question of why there is no Nobel Prize in mathematics is a matter of some speculation; a popular but probably apocryphal anecdote has it that at the time the Nobel Prizes were endowed, Alfred Nobel's wife was having an affair with Gustav Mittag-Leffler, a distinguished Swedish mathematician. Yes, mathematics has its Fields Medal, awarded every four years, but it is awarded only to mathematicians under forty. If you win it you are set for life, prestige-wise, but you're not going to be able to put your kids through college on the proceeds.

At the turn of the millennium, the Clay Mathematics Institute posted seven critical problems in mathematics—and offered an unheard-of $1 million for the solution of each. Some of the problems, such as the Birch and Swinnerton-Dyer conjecture, are highly technical and even the statement of the problem is comprehensible only to specialists in the field. Two of these problems, the Navier-Stokes equation and the Yang-Mills theory, are in the realm of mathematical physics. Solutions to these problems will enable a better understanding of the physical universe, and may actually enable significant technological advances. One of these problems, however, is related to one of the most mystifying of life's little annoyances: Why is your car never ready at the time the garage promised it?

Putting a Man on the Moon

When President John F. Kennedy promised that America would put a man on the moon by the end of the 1960s, he almost certainly did not foresee many of the numerous side effects that the space race would produce. Of course, the space race gave the microelectronic industry a huge boost, leading to calculators and personal computers. Two lesser results were Tang, an orange-flavored drink for the astronauts that would soon be found on supermarket shelves, and Teflon, a superslick material that would not only be used as a coating for numerous cooking devices, but would also insinuate itself into the English language as a synonym for a politician to whom charges of malfeasance would not adhere. Finally, the space race resulted in a series of insights as to why the world never seems to function as well as it should.

America had previously engaged in one other mammoth technological undertaking, the Manhattan Project, but developing the atomic bomb was relatively simple when compared to the problem of putting a man on the moon—at least from the standpoint of scheduling. There were three major components of the Manhattan Project—bomb design and testing, uranium production, and mission training. The first two could proceed independently, although actual testing awaited the arrival of sufficient fissionable material from factories at places such as Hanford and Oak Ridge. Mission training began only when the specifications for the weapon were reasonably well known, and was relatively simple—make certain there was a plane that could carry it and a crew that could fly it.

From the standpoint of scheduling, putting a man on the moon was a far more difficult task. There was a tremendous amount of coordination needed between the industrial complex, the scientific branch, and the astronaut training program. Even as apparently simple a matter as planning the individual mission responsibilities of the lunar astronauts had to be carefully choreographed. In sending astronauts to the moon, a lot of tasks had to be precisely scheduled so as to make optimal use of the available time while simultaneously making sure that outside constraints were also satisfied—such as making sure the space capsule was rotated so it did not overheat. Thus was born the branch of mathematics known as scheduling, and with it the discovery of how improving the individual components that go into an ensemble can result in counterproductive—and counterintuitive—outcomes.

So Why Is Your Car Never Ready When They Promised?

Whether your neighborhood garage is in Dallas, Denver, or Des Moines, it encounters basically the same problem. On any given day, there are a bunch of cars that need work done, and equipment and mechanics available to do the job. If only one car comes into the shop, there's no problem with scheduling, but if several cars need repairs, it's important to do things efficiently. There may be only one diagnostic analyzer and only two hydraulic lifts—ideally, one would want to schedule the repair sequence so that everything is simultaneously in operation, as idle time costs money. The same thing can be said about the available mechanics; they're paid by the hour, so if they are sitting on the sidelines while cars are waiting to be serviced, that costs money, too.

One critical aspect of scheduling is a method of displaying the tasks to be done, how they relate to one another, and how long they will take. For instance, to determine if a tire has a leak, the tire must be removed before checking it in the water bath. The standard way of displaying the tasks, their times, and their relationships to each other is by means of a digraph. A digraph is a diagram with squares and arrows indicating the tasks to be done, the order in which they are to be done, and the time required—one such is indicated below.

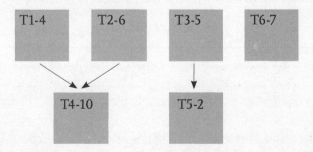

Task 1 requires 4 time units (hours, days, months—whatever), and tasks 1 and 2 must be completed before task 4, which requires 10 time units, is undertaken. Similarly, task 3 must be completed before task 5 is begun.

Finally, task 6 can be done at any time—nothing needs to be done before it, and it is not a prerequisite for any other task. Additionally, each task must be assigned to a single worker and not broken up into subtasks—if we could do this, we'd simply label each subtask as a separate task.

A little additional terminology is associated with the above digraph. A task is ready if all prerequisites for the task have been completed. In the above diagram, tasks 1, 2, 3, and 6 are ready at the outset, whereas task 5 will be ready only when task 3 has been completed, and task 4 will be ready only when both tasks 1 and 2 are completed. Notice that it will take a minimum of 16 time units to complete all the tasks, as task 2 followed by task 4, which requires 16 time units, is the critical path—the path of longest duration.

Numerous algorithms have been devised for scheduling tasks; we'll examine just one of them, which is known as priority-list scheduling. The idea is simple. We make a list of the tasks in order of importance. When a task is finished, we cross it off the list. If someone is free to work on a task, we set that person to work on the most important unfinished task, as determined by the priority list—if several mechanics are free, we assign them in alphabetical order. The algorithm does not describe how the priority list is constructed—for instance, if the garage owner's wife needs her oil changed, that item may be placed at the top of the priority list, and if someone slips the garage owner $20 for extra-fast service, that might go right behind it.

To illustrate how all this stuff comes together, let's suppose that times in the above digraph are measured in hours, and our priority list is T1, T2, T4, T3, T5, T6. If Al is the only mechanic on hand, there is no real scheduling to be done—Al just does all the jobs on the priority list in that order, and it takes him a total of 32 hours (the sum of all the times) to finish all the tasks. However, if the garage hires Bob, another mechanic, we use the priority list to construct the following schedule.

Mechanic	Task Start Times						
	0	4	6	9	11	16	18
Al	T1	T3		T5	T6		Done
Bob	T2		T4			Idle	Done

Since tasks 1 and 2 are at the head of the priority list and both are ready at the start, we schedule Al for task 1 and Bob for task 2. When Al finishes task 1, at the end of 4 hours, the next task on the priority list is task 4—but task 4 isn't ready yet, as Bob hasn't finished task 2. So Al must bypass task 4 and start in on task 3, the next task on the priority list. The rest of the diagram is pretty straightforward. This schedule is as good as we could hope for, as there are a total of 34 hours that must be scheduled, and there is no way we can schedule 17 hours for each mechanic (unless we allow a task to be broken up between two mechanics, which isn't allowed by the rules). It finishes all tasks as quickly as possible, and minimizes the amount of idle time, two frequently used criteria in constructing schedules.

When Making Things Better Actually Makes Things Worse

The interaction between the task digraph and the priority list is complicated, and unexpected situations can arise.

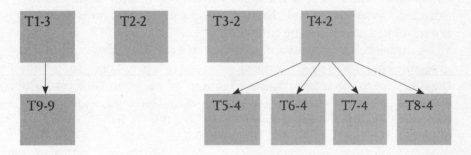

The priority list is just the tasks in numerical order: T1, T2, T3, ..., T9. The garage has three mechanics: Al, Bob, and Chuck. The schedule that results appears below.

Mechanic	Task Start Times					
	0	2	3	4	8	12
Al	T1		T9			Done
Bob	T2	T4		T5	T7	Done
Chuck	T3	Idle		T6	T8	Done

From a schedule standpoint, this is a "perfect storm" scenario. The critical path is 12 hours long, all tasks are finished by this time, and we have minimized the amount of idle time, as there are 34 hours of tasks to be done and three mechanics available for 12 hours would be a total of 36 hours.

If the garage has a lot of business, it might decide to hire an extra mechanic. If the jobs to be done conform to the above digraph and the same priority list, we would certainly expect that there would be a lot more idle time, but the resulting schedule contains a surprise.

Mechanic	Task Start Times					
	0	2	3	6	7	15
Al	T1		T8		Idle	Done
Bob	T2	T5		T9		Done
Chuck	T3	T6		Idle		Done
Don	T4	T7		Idle		Done

A postmortem on this schedule reveals that the trouble started when Don was assigned task 4 at the start. This made task 8 available "too early," and so Al can take it on, with the result that task 9 gets started 3 hours later than in the original schedule. This is certainly somewhat unexpected, as you would think that having more mechanics available would not result in a later finishing time.

The garage has an alternative to hiring an extra mechanic—it can upgrade the equipment used for the various tasks. When it does so, it finds that the time for each task has been reduced by one hour. We'd certainly expect good things to happen after the upgrade. The original job digraph is now modified to the one below.

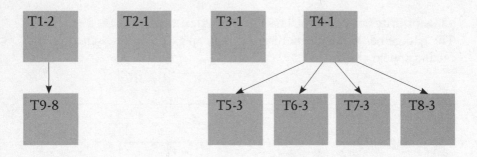

When the same priority list (and, of course, the priority-list scheduling algorithm) is used for the original three mechanics, the following schedule gets constructed.

Mechanic	Task Start Times					
	0	1	2	5	8	13
Al	T1		T5	T8	Idle	Done
Bob	T2	T4	T6	T9		Done
Chuck	T3	Idle	T7	Idle		Done

This schedule could well be the poster child for the how-making-everything-better-sometimes-makes-things-worse phenomenon. Improving the equipment reduced the length of the critical path, but actually slowed things down, rather than speeding things up! Yes, there are lots of other scheduling algorithms available, but the magic bullet has yet to be found—no algorithm yet studied has generated consistently optimal schedules. What is worse, there may be no such algorithm—at least, not one that can be executed in a reasonable period of time!

However, there is one such algorithm that always works—construct all possible schedules that satisfy the digraph, and choose the one that best optimizes whatever criteria are used. There's a major problem with that: there could be an awful lot of schedules, especially if there are a lot of tasks. We shall examine this situation in more depth in chapter 9, when we discuss what is known in mathematics as the P versus NP problem.

The Short-Order Cook, Two Georges, and Moneyball

When I was in graduate school, I would occasionally splurge by going out for breakfast. The diner I frequented was typical for the 1960s—a few tables and a Formica counter with individual plastic seats surrounding a large rectangular grill on which the short-order cook could be seen preparing the orders. The waitresses would clip the orders to a metal cylinder, and when he had a free moment the cook would grab them off the cylinder and start preparing them.

This particular cook moved more gracefully than anyone you are likely to see on *Dancing with the Stars*. When sections of the grill dried out or became covered with the charred remainders of eggs or hashed browns, he scraped them off and poured on a thin layer of oil. Eggs were cooked on one quadrant of the grill, pancakes and French toast on a second, hashed browns on a third, and bacon and ham on the fourth. He never seemed hurried, always arriving just in time to flip over an egg that had been ordered over easy, or to prevent bacon or hashed browns from burning. Some people find fascination in watching construction workers, but I'll take a good short-order cook over construction workers any time.

There is a certain poetry to the smooth integration of an assortment of tasks that is sought in practically every enterprise that requires such an integration—but how best to accomplish it? A notable arena for such endeavors is professional sports, in which team chemistry, the melding of accomplished individuals into a cohesive unit, is the ultimate goal. Oft-tried algorithms have decidedly mixed results. One such algorithm could be described as "buy the best." Jack Kent Cooke hired George Allen to coach the Washington Redskins, and said of him that "I gave George an unlimited budget, and he overspent it." George Steinbrenner, the owner of the New York Yankees, is a firm believer in the theory that if one pays top dollar for top professionals, one produces top teams. The payroll for the New York Yankees in 2006 exceeded $200 million—and while the team got to the play-offs, they lost to the Detroit Tigers in the first round, an event cheered not only by Tiger fans but by confirmed Yankee haters such as myself.

On the other side of the algorithm divide is the belief that if one tries to buy components by minimizing the dollars-spent-per-past-desirable-outcome-achieved (such as purchasing a cleanup hitter using dollars per home run hit last year), good results can be obtained with a limited budget. This approach, known as "moneyball," was developed by Billy Beane, the general manager of the Oakland Athletics, who constructed several remarkably successful teams while spending very little money. One of his disciples was Paul DePodesta, who took over my beloved Los Angeles Dodgers (actually, I'm a Cub fan, but the Dodgers are beloved of my wife, and when the woman is happy the man is happy)—and ruined them with the moneyball philosophy. DePodesta was summarily dismissed and replaced by Nick Colletti, a man with a solid baseball pedigree, and the Dodgers have made it back to the play-offs twice in the last four years.

While the examples cited above come from professional sports, the goals of any organization are similar. If the magic formula for organizational success in professional sports is discovered, you can bet the farm that management experts will study this formula in order to adopt it to other enterprises. Today the Dodgers, tomorrow Microsoft.

So what's the lesson? The lesson, which we shall investigate more thoroughly later in the book, is that some problems may well be so complex that there is no perfect way to solve them.

Unless you are a professional mathematician, you have no chance of coming up with a solution to the Birch and Swinnerton-Dyer conjecture, but any person of reasonable intelligence can probably devise a variety of scheduling algorithms. Want to take a shot? One of the attractive aspects to a mathematical problem is that the only items needed are paper, pencil, and time—but be aware that this problem has resisted the best efforts of several generations of mathematicians.

Mathematics and science have stood at the threshold of great unsolved problems before. Two millennia of mathematicians had worked arduously to discover the solutions of polynomial equations of degree four or less, and in the sixteenth century the general solution of the quintic (the polynomial of degree five) was the goal of the best algebraists in the world. The physics community was likewise poised at the turn of the twentieth century, seeking a way out of the ultraviolet catastrophe—the prediction that a perfectly black object in thermal equilibrium would emit radiation with infinite power.

An equally challenging puzzle confronted social scientists a relatively short while ago. The dictatorships that had strangled Germany, Italy, and Japan had been overthrown as a result of the Second World War. With democracies emerging throughout the world, the social scientists of the day were eagerly continuing a quest begun two centuries previously, the search for the ideal method of translating the votes of individuals into the wishes of the society.

All these efforts would lead to related dramatic discoveries—that there are some things we cannot know, some things we cannot do, and some goals we cannot achieve. Possibly some mathematician will pick up the Clay millennium jackpot by discovering that there is no perfect way to create schedules, and we'll just have to resign ourselves to hearing that our car isn't ready when we call the garage to inquire if we can pick it up.

Describing the Universe

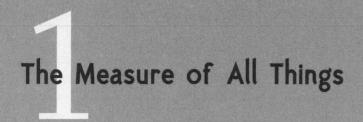

1 The Measure of All Things

Missed It by THAT Much

According to Plato, Protagoras was the first sophist, or teacher of virtue—a subject that greatly fascinated the Greek philosophers. His most famous saying was "Man is the measure of all things: of things which are, that they are, and of things which are not, that they are not."[1] The second part of the sentence established Protagoras as the first relativist, but to me the first part of the sentence is the more interesting, because I think Protagoras missed it by just a single letter. Things have their measure—it is an intrinsic property of things. Man is not the measure of all things, but the measurer of all things.

Measurement is one of man's greatest achievements. While language and tools may be the inventions that initially enabled civilization to exist, without measurement it could not have progressed very far. Measurement and counting, the obvious predecessors to measurement, were man's initial forays into mathematics and science. Today, Protagoras's statement still raises questions of profound interest: How do we measure things that are, and can we measure things that are not?

What Is This Thing Called Three?

Math teachers in college generally teach two different types of classes: classes in which relatively high-level material is taught to students who will use it in their careers, and classes in which relatively low-level material is taught to students who, given the choice of taking the class or a root canal without anesthesia, might well opt for the latter. The second type of class includes the math courses required by the business school—most of the students in these classes believe they will someday be CEOs, and in the unlikely event they ever need a math question answered they will hire some nerd to do it. It also includes math for liberal arts students, many of whom believe that the primary use for numbers are labels—such as "I wear size 8 shoes"—and the world would function better if different labels, such as celebrities or cities, were used instead. After all, it might be easier to remember that you wear Elvis shoes or Denver shoes than to remember that you wear size 8 shoes. Don't laugh—Honda makes Accords and Civics, not Honda Model 1 and Honda Model 2.

Fortunately (for at my school all teachers frequently teach lower-level courses), the second type of math class also includes my favorite group of students—the prospective elementary school teachers, who will take two semesters of math for elementary school teachers. I have the utmost respect for these students, who are planning on becoming teachers because they love children and want to make life better for them. They're certainly not in it for the money (there's not a whole lot of that), or for the freedom from aggravation (they frequently have to teach in unpleasant surroundings with inadequate equipment, indifferent administrators, hostile parents, and all sorts of critics from politicians to the media).

Most of the students in math for elementary school teachers are apprehensive on the first day of class—math generally wasn't their best subject, and it's been a while since they've looked at it. I believe that students do better if they are in a comfortable frame of mind, so I usually start off with Einstein's famous quote, "Do not worry about your difficulties with mathematics; I assure you mine are far greater."[2] I then proceed to tell them that I've been teaching and studying math for half a century, and they know just about as much about "three" as I do—because I can't even tell them what "three" is.

Sure, I can identify a whole bunch of "threes"—three oranges, three cookies, etc.—and I can perform a bunch of manipulations with "three" such as two plus three is five. I also tell them that one of the reasons that mathematics is so useful is because we can use the statement "two plus three is five" in many different situations, such as knowing we'll need $5 (or a credit card) when we purchase a muffin for $2 and a frappuccino for $3. Nonetheless, "three" is like pornography—we know it when we see it, but damned if we can come up with a great definition of it.

More, Less, and the Same

How do you teach a child what a tree is? You certainly wouldn't start with a biologist's definition of a tree—you'd simply take the child out to a park or a forest and start pointing out a bunch of trees (city dwellers can use books or computers for pictures of trees). Similarly with "three"—you show the child examples of threes, such as three cookies and three stars. In talking about trees, you would undoubtedly point out common aspects—trunks, branches, and leaves. When talking about threes to children, we make them do one-to-one matching. On one side of the page are three cookies; on the other side, three stars. The child draws lines connecting each cookie to a different star; after each cookie has been matched to different stars, there are no unmatched stars, so there are the same number of cookies as stars. If there were more stars than cookies, there would be unmatched stars. If there were fewer stars than cookies, you'd run out of stars before you matched up all the cookies.

One-to-one matching also reveals a very important property of finite sets: no finite set can be matched one-to-one with a proper subset of itself (a proper subset consists of some, but not all, of the things in the original set). If you have seventeen cookies, you cannot match them one-to-one with any lesser number of cookies.

The Set of Positive Integers

The positive integers 1, 2, 3, . . . are the foundation of counting and arithmetic. Many children find counting an entertaining process in itself, and sooner or later stumble upon the following question: Is there a largest number? They can generally answer this for themselves—if there were a largest number of cookies, their mother could always bake another one. So there is no number (positive integer) that describes how many numbers (positive integers) there are. However, is it possible to come up with something that we can use to describe how many positive integers there are?

There is—it's one of the great discoveries of nineteenth-century mathematics, and is called the cardinal number of a set. When that set is finite, it's just the usual thing—the number of items in the set. The cardinal number of a finite set has two important properties, which we discussed in the last section. First, any two sets with the same finite cardinal number can be placed in one-to-one correspondence with each other; just as a child matches a set of three stars with a set of three cookies. Second, a finite set cannot be matched one-to-one with a set of lesser cardinality—and in particular, it cannot be matched one-to-one with a proper subset of itself. If a child starts with three cookies, and eats one, the remaining two cookies cannot be matched one-to-one with the original three cookies.

Hilbert's Hotel

The German mathematician David Hilbert devised an interesting way of illustrating that the set of all integers can be matched one-to-one with a proper subset of itself. He imagined a hotel with an infinite number of rooms—numbered R1, R2, R3, ... The hotel was full when an infinite number of new guests, numbered G1, G2, G3, . . . arrived, requesting accommodations. Not willing to turn away such a profitable source of revenue, and being willing to discomfit the existing guests to some extent, the proprietor moved the guest from R1 into R2, the guest from R2 into R4, the guest from R3 into R6, and so on—moving each guest into a new room with twice the room number of his or her current room. At the end of this procedure, all the even-numbered rooms were occupied, and all the odd-numbered rooms were vacant. The proprietor then moved guest G1 into vacant room R1, guest G2 into vacant room R3, guest G3 into vacant room R5, ... Unlike every hostelry on planet Earth, Hilbert's Hotel never has to hang out the No Vacancy sign.

In the above paragraph, by transferring the guest in room N to room 2N, we have constructed a one-to-one correspondence between the positive integers and the even positive integers. Every positive integer is matched with an even positive integer, via the correspondence $N \leftrightarrow 2N$, every even positive integer is matched with a positive integer, and different integers are matched with different even positive integers. We have matched an infinite set, the positive integers, in one-to-one fashion with a proper subset, the even positive integers. In doing so, we see that infinite sets differ in a significant way from finite sets—in fact, what distinguishes infinite sets from finite sets is that infinite sets can be matched one-to-one with proper subsets, but finite sets cannot.

Ponzylvania

There are all sorts of intriguing situations that arise with infinite sets. Charles Ponzi was a swindler in early twentieth-century America who devised plans (now known as Ponzi schemes) for persuading people to invest money with him for a significant return. Ponzi schemes are highly pernicious (which is why they're illegal)—periodically, the country is inundated with a new version, such as pyramid investment clubs.[3] Ponzi paid early investors with the funds from investors who anted up later, creating the impression that his investors were prospering—at least, the early ones. The last ones to invest were left holding the bag, as it is impossible to continue paying profits to investors by this method unless later investors are found—and, eventually we run out of later investors. Everywhere, that is, but in Ponzylvania.

B.P. (Before Ponzi), Ponzylvania was a densely populated country that had incurred overwhelming debt. Its inhabitants, like the rooms in Hilbert's Hotel, are infinite in number—we'll call them $I1, I2, I3, \ldots$. Every tenth inhabitant ($I10, I20, \ldots$) has a net worth of $1, while all the others are $1 in debt. The total assets of inhabitants 1 through 10 is therefore minus $9, as are the total assets of inhabitants 11 through 20, 21 through 30, and so on. Every group of 10 successively numbered inhabitants has negative total assets.

Not to worry; all that is needed is a good way of rearranging assets, so enter Charles Ponzi—a criminal in the United States, but a national hero in Ponzylvania. He collects a dollar from $I10$ and a dollar from $I20$, giving them to $I1$, who now has a net worth of $1. He then collects a dollar from $I30$ and a dollar from $I40$, giving them to $I2$, who also now has a net worth of $1. He then collects a dollar from $I50$ and a dollar from $I60$, giving them to $I3$, who also now has a net worth of $1. We'll assume that when he comes to an inhabitant such as $I10$, who is now flat broke (he originally had a dollar, but it was given to $I1$ early on), he simply transfers a dollar from the next untapped dollar owner. He continues this process until he has gone through all the inhabitants—at the end of which everyone has $1!

You don't become a national hero by giving everyone assets of a dollar—so Ponzi embarks upon Stage 2 of his master financial plan. Since everyone has a dollar, he collects the dollars from $I2, I4, I6, I8, \ldots$ and gives them to $I1$. $I1$, now infinitely wealthy, retires to his seaside villa. This process leaves $I3, I5, I7, I9, \ldots$ with $1 each. The key point here is that there are still infinitely many inhabitants, each of whom has a dollar. Ponzi now collects the dollars from $I3, I7, I11, I15$ (every other odd number), . . . and gives them to $I2$, who also retires to his seaside villa. At this juncture, there are *still* infinitely many inhabitants who have a dollar ($I5, I9, I13, \ldots$), so Ponzi collects a dollar from every other dollar-owning inhabitant ($I5, I13, I21, \ldots$) and gives them to $I3$. At the end of this process, $I3$ retires to his seaside villa, and there are *still* infinitely many inhabitants who have a dollar. At the end of Stage 2, everyone is enjoying life on his or her seaside villa. No wonder they renamed the country in his honor.

The intellectual resolution of this particular Ponzi scheme involves rearrangements of infinite series, a topic generally not covered until a math major takes a course in real analysis. Suffice it to say that there are problems, which go to the heart of how infinite arithmetic processes differ from finite ones—when we tallied the total assets of the country by looking at the total assets of I1 through I10 (minus \$9) and adding them to the total assets of I11 through I20 (minus \$9), and so on, we get a different outcome from when we total the assets by adding $(I10 + I20 + I1) + (I30 + I40 + I2) + (I50 + I60 + I3) + \ldots = (1 + 1 + -1) + (1 + 1 + -1) + (1 + 1 + -1) + \ldots = 1 + 1 + 1 + \ldots$. The two different ways of collecting money (doing arithmetic) yield different results. Unlike bookkeeping in the real world, in which no matter how you rearrange assets the total is always the same, a good bookkeeper in Ponzylvania can spin gold from straw.

Georg Cantor (1845–1918)

Until Georg Cantor, mathematicians had never conducted a successful assault on the nature of infinity. In fact, they hadn't really tried—so great a mathematician as Carl Friedrich Gauss had once declared that infinity, in mathematics, could never describe a completed quantity, and was only a manner of speaking. Gauss meant that infinity could be approached by going through larger and larger numbers, but was not to be viewed as a viable mathematical entity in its own right.

Perhaps Cantor's interest in the infinite might have been predicted, given his unusual upbringing—he was born a Jew, converted to Protestantism, and married a Roman Catholic. Additionally, there was a substantial amount of artistic talent in the family, as several family members played in major orchestras, and Cantor himself left a number of drawings that were sufficient to show that he possessed artistic talent as well.

Cantor took his degree in mathematics under the noted analyst Karl Theodor Wilhelm Weierstrass, and Cantor's early work traveled along the path marked out by his thesis adviser—a common trait among mathematicians. However, Cantor's interest in the nature of infinity persuaded him to study this topic intensely. His work generated considerable interest in the mathematical community—as well as considerable controversy. Cantor's work flew in the face of Gauss, as it dealt with infinities as completed quantities in a manner analogous to finite ones.

Among the mathematicians who had a great deal of difficulty accepting this viewpoint was Leopold Kronecker, a talented but autocratic German mathematician. Kronecker exerted his influence from his chair at the prestigious University of Berlin, while Cantor was relegated to the minor leagues at the University of Halle. Kronecker was an old-school mathematician who took Gauss at his word on the subject of infinity, and he did his best to denigrate Cantor's works. This helped spark numerous outbreaks of depression and paranoia in Cantor, who spent much of his later life in mental institutions. It did not help matters that Cantor proclaimed his mathematics to be a message from God, and that his other interests included attempting to convince the world that Francis Bacon wrote the works of Shakespeare.

Nonetheless, between periods of confinement, Cantor produced works of stunning brilliance, results which changed the direction of mathematics. Sadly, he died in the mental institution where he had spent much of his adult life. Just as the greatness of Mozart and van Gogh became apparent after their deaths, so did the work of Cantor. Hilbert described transfinite arithmetic, one of Cantor's contributions, as "the most astonishing product of mathematical thought, one of the most beautiful realizations of human activity in the domain of the purely intelligible."[4] Hilbert continued by declaring that "No one shall expel us from the paradise which Cantor has created for us."[5] One can only wonder how Cantor's life would have differed if Hilbert, rather than Kronecker, had been the one holding down the chair at the University of Berlin.

Another Visit to Hilbert's Hotel

One of Cantor's great discoveries was that there were infinite sets whose cardinality was greater than that of the positive integers—infinite sets that could not be matched one-to-one with the positive integers. Such a set is the collection of all people with infinitely long names.

An infinitely long name is a sequence of letters A through Z and blanks—one letter or blank for each of the positive integers. Some people, such as "Georg Cantor," have names consisting mostly of blanks—the first letter is a G, the second letter an E, . . . , the sixth letter a blank, the twelfth letter an R, and letters thirteen, fourteen, . . . (the three dots stand for "on and on forever," or some such phrase) are all blanks. Some people, such as "AAAAAAAAA . . . ," have names consisting exclusively of letters—every letter of her name is an A. Of course, it takes her quite a while to fill out the registration card at Hilbert's Hotel, but we'll dispense with that problem for the time being.

The collection of all people with infinitely long names cannot be matched one-to-one with the integers. To see that this is the case, assume that it could be so matched. If so, then every person with an infinitely long name could be assigned a room in Hilbert's Hotel, and we'll assume we've done so. We'll demonstrate a contradiction by showing that there is a person with an infinitely long name who has no room in the hotel.

To do this, we'll construct the name of such a person, whom we'll call the mystery guest, letter by letter. Look at the name of the person in room R1, and choose a letter different from the first letter of that name. That "different letter" is the first letter of our mystery guest's name. Then look at the name of the person in room R2, and choose a letter different from the second letter of that name. That "different letter" is the second letter of our mystery guest's name. In general, we look at the nth letter of the name of the guest in room Rn, and choose a 'different letter' from that one as the nth letter of our mystery guest's name.

So constructed, our mystery guest is indeed roomless. He's not in R1, because the first letter of his name differs from the first letter of the guest in R1. Our guest is not in R2, because the second letter of his name differs from the second letter of the guest in R2. And so on. Our mystery guest is nowhere to be found in Hilbert's Hotel, and so the collection of people with infinitely long names cannot be matched one-to-one with the positive integers.

Great results in mathematics have the name of their discoverer attached, such as the Pythagorean theorem. Mathematical objects worthy of study have the name of an important contributor affixed, such as "Cantor set." Brilliant mathematical proof techniques are similarly immortalized—the above construction is known as a "Cantor diagonal proof" (if we were to arrange the names of the hotel guests in a list from top to bottom, with the first letters of each name comprising the first column, the second letters of each name comprising the second column, and so on, the line connecting the first letter of the first name to the second letter of the second name, thence to the third letter of the third name, and so on, would form the diagonal of the infinite square that comprises the list). In fact, Cantor is one of the few mathematicians to hit for the cycle, having not only proof techniques named for him, but theorems and mathematical objects as well.

The Continuum Hypothesis

It is fairly easy to see that the above proof technique shows that the collection of real numbers between 0 and 1 also has a different cardinal number than that of the positive integers. The real numbers between 0 and 1 (known as "the continuum") are, when written in decimal expansion, simply infinitely long names with letters 1 through 9 rather than A through Z and 0 instead of blank. For example, ¼ = .25000. . . . Cantor worked out the arithmetic of cardinal numbers, and designated the cardinal number of the positive integers as aleph-0, and the cardinal number of the continuum as c.

A lot of mileage can be gained from the Cantor diagonal proof. Cantor used it to show that the set of rational numbers has cardinality aleph-0, as does the set of algebraic numbers (all those numbers that are roots of polynomials with integer coefficients). Also, it can be used to show the infinite analogy of the child's result that there is no largest (finite) number. Cantor was able to show that, for any set S, the set of all subsets of S could not be matched one-to-one with the set S, and so had a larger cardinal number. As a result, there is no largest cardinal number.

Filling the Gaps

Leopold Kronecker, when he wasn't making life miserable for Cantor, was a mathematician of considerable talent, and is also the author of one of the more famous quotations in mathematics: "God made the integers, all else is the work of man."[6] One of the first jobs that man had to do was fill in the gaps between the integers in the number line. The task of filling the gaps was to return in the nineteenth century, when mathematicians encountered the problem of whether there existed cardinal numbers between aleph-0 and c. As explained above, efforts to show that obvious sets, such as the set of rational numbers and the set of algebraic numbers, had different cardinal numbers from aleph-0 and c proved unsuccessful. Cantor hypothesized that there was no such cardinal number—every subset of the continuum had cardinality aleph-0 or c; this conjecture became known as the continuum hypothesis. Proving or disproving the continuum hypothesis was a high priority for the mathematical community. In a key turn-of-the-century mathematics conference, David Hilbert listed the solution of this problem as the first on his famous list of twenty-three problems that would confront mathematicians in the twentieth century. Solution of just one of these problems would make the career of any mathematician.

The Axiom of Choice

The axiom of choice is a relatively recent arrival on the mathematical scene—in fact, it wasn't until Cantor arrived on the mathematical scene that anybody even thought that such an axiom was necessary. The axiom of choice is simple to state; it says that if we have a collection of nonempty sets, we can choose a member of each set. In fact, when I first saw this axiom, my initial reaction was "Why do we need this axiom? Choosing things from sets is like shopping with an inexhaustible budget. Just go into a store [set], and say, 'I want this.'" Nonetheless, the axiom of choice is highly controversial—insofar as an axiom could ever be considered controversial.

The controversy centers around the word *choose*. Just as there are activist judges and strict constructionists, there are liberal mathematicians and strict constructionists when it comes to the word *choose*. Is choice an active process, in which one must specify the choices made (or a procedure for making those choices), or is it merely a statement of existence, in that choices can be made (this is somewhat reminiscent of Henry Kissinger's remark that "mistakes were made in Administrations of which I was a part")?[7] If you are a strict constructionist who wants a recipe for choice, you won't have any problem doing this with a collection of sets of positive integers—you could just choose the smallest integer in any set. In fact, there are many collections of sets in which constructing a choice function (a function whose value for each set is the choice that is made for that set) presents no problem. However, if one considers the collection of all nonempty subsets of the real line, there is no obvious way to do this—nor is there an unobvious way, as no one has yet done it and the betting of many mathematical logicians is that it can't be done.

There is a significant difference between "sets of positive integers" and "sets of real numbers"—and that is the existence of a smallest positive integer in any nonempty set of positive integers—but there is no obvious smallest real number in any nonempty set of real numbers. If there were, we could find a choice function in exactly the same manner that we did for sets of positive integers—we'd simply choose the smallest real number in the nonempty set.

It may have occurred to you that there are sets of real numbers that clearly have no smallest member, such as the set of all positive real numbers. If you think you have the smallest such number, half of it is still positive, but smaller. However, there might conceivably be a way to arrange the real numbers in a different order than the usual one, but one such that every nonempty set of real numbers has a smallest member. If there were, then the choice function would be the one defined in the last paragraph—the smallest number in each set. As a matter of fact, this idea is known as the well-ordering principle, and is logically equivalent to the axiom of choice.

If finding a choice function for the collection of all subsets of real numbers gives you a headache, you might prefer the following version of the dilemma, due to Bertrand Russell—if you have an infinite number of pairs of shoes, it is easy to choose one shoe from each pair (you could choose the left shoe), but if you have an infinite number of pairs of socks, there is no way to distinguish one sock from another, and so you can't explicitly write out a way to choose one from each pair.

The great majority of mathematicians favor the existence formulation—a choice exists (possibly in some abstract never-neverland in which we cannot specify how), and an incredible amount of fascinating mathematics has resulted from incorporating the axiom of choice. Far and away the most intriguing of the results is the Banach-Tarski paradox,[8] the statement of which usually results in people feeling that mathematicians have lost their collective minds. This theorem states that it is possible to decompose a three-dimensional sphere into a finite number of pieces and rearrange them by rotations and translations (moving from one point of space to another by pushing or pulling, but not rotating) into a sphere with twice the radius of the original. Tempting though it may be to buy a small golden sphere for a few hundred bucks, Banach-Tarskify it to double its radius, and do so repeatedly until you have enough gold to retire to *your* seaside villa, not even Charles Ponzi can help you with this one. Unfortunately, the pieces into which the sphere can be decomposed (notice that I did not use the word *cut*, which is an actual physical process), exist only in the abstract never-neverland of what are called "nonmeasurable sets." No one has ever seen a nonmeasurable set and no one ever will—if you can make it, then it is not nonmeasurable, but if you accept the axiom of choice in the existence sense, there is an abundance of these sets in that never-neverland.

Consistent Sets of Axioms

I'm not sure that other mathematicians would agree with me, but I think of mathematicians as those who make deductions *from* sets of axioms, and mathematical logicians as those who make deductions *about* sets of axioms. On one point, though, mathematicians and mathematical logicians are in agreement—a set of axioms from which contradictory results can be deduced is a bad set of axioms. A set of axioms from which no contradictory results can be deduced is called consistent. Mathematicians generally work with axiom sets that the community feels are consistent (even though this may not have been proven), whereas among the goals of the mathematical logicians are to prove that axiom sets are consistent.

Just as there are different geometries (Euclidean, projective, spherical, hyperbolic—to name but a few), there are different set theories. One of the most widely studied is the axiomatic scheme proposed by Ernst Zermelo and Adolf Fraenkel, who came up with a system to which was added the axiom of choice.[9] The industry-standard version of set theory is known as ZFC—the Z and F stand for you-know-who, and the C for the axiom of choice. Mathematicians are inordinately fond of abbreviations, as the mathematical aesthetic dictates that conveying a lot of meaning in very few symbols is attractive, and so CH is the abbreviation for the continuum hypothesis.

The first significant dent in Hilbert's first problem was made in 1940 by Kurt Gödel (of whom we shall hear much more in a later chapter), who showed that if the axioms of ZFC were consistent, then including CH as an additional axiom to produce a larger system of axioms, denoted ZFC+CH, did not result in any contradictions, either.

This brought the continuum hypothesis, which had been under scrutiny by mathematicians (who would have liked either to find a set of real numbers with a cardinal number other than aleph-0 or c, or prove that such a set could not exist), into the realm of mathematical logic. In the early 1960s, Paul Cohen of Stanford University shocked the mathematical community with two epic results. He showed that if ZFC were consistent, CH was undecidable within that system; that is, the truth of CH could not be determined using the logic and axioms of ZFC. Cohen also showed that including the negation of CH (abbreviated "not CH") to ZFC to produce the system ZFC+not CH was also consistent. In conjunction with Gödel's earlier result, this showed that it didn't matter whether you assumed CH was true or CH was false, adding it to an assumed-to-be-consistent ZFC produced a theory that was also consistent. In the language of mathematical logic, CH was independent of ZFC. This work was deemed so significant that Cohen (who passed away in the spring of 2007), was awarded a Fields Medal in 1966.

What did this mean? One way to think of it is to hark back to another situation in which an important hypothesis proved to be independent of a prevailing set of axioms. When Euclidean geometry was subjected to investigation, it was realized that the parallel postulate (through each point not on a given line l, one and only one line parallel to l can be drawn) was independent of the other axioms. Standard plane geometry incorporates the parallel postulate, but there exist other geometries in which the parallel postulate is false—in hyperbolic geometry, there are at least two lines that can be drawn through any point off the line l that are parallel to l. Logicians say that plane geometry is a model that incorporates the parallel postulate, and hyperbolic geometry is a model that incorporates the negation of the parallel postulate.

The Continuum: Where Are We Now?

One of the preeminent physicists of today, John Archibald Wheeler (whom we shall encounter when we discuss quantum mechanics), feels that both the discrete structure of the integers and the fundamental nature of the continuum are vital to the work of physics, and weighs in with a physicist's point of view.

> For the advancing army of physics, battling for many a decade with heat and sound, fields and particles, gravitation and space-time geometry, the cavalry of mathematics, galloping out ahead, provided what it thought to be the rationale for the real number system. Encounter with the quantum has taught us, however, that we acquire our knowledge in bits; that the continuum is forever beyond our reach. Yet for daily work the concept of the continuum has been and will continue to be as indispensable for physics as it is for mathematics. In either field of endeavor, in any given enterprise, we can adopt the continuum and give up absolute logical rigor, or adopt rigor and give up the continuum, but we can't pursue both approaches at the same time in the same application.[10]

Wheeler sees a clash between the current quantum view of reality (Wheeler's absolute logical rigor) and the continuum, a useful mathematical idealization that can never be. Mathematicians are lucky—they do not have to decide whether the object of their investigation is either useful or a great description of reality. They merely have to decide if it is interesting.

Given Cohen's result on the undecidability of CH within ZFC, and since CH is independent of ZFC, what are the choices for continuing research? The problem has basically been removed from the domain of the mathematician, most of whom are content with ZFC as an axiomatic framework. The majority of logicians concentrate on the ZFC part of the problem, and much work is being done on constructing other axioms for set theory in which CH is true. Future generations of mathematicians may well decide to change the industry standard, and abandon ZFC for some other system.

Of what value is all this? From the mathematical standpoint, even though developments in the twentieth century have diminished the importance of solving Hilbert's first problem, the continuum is one of the fundamental mathematical objects—added knowledge of its structure is of paramount importance, just as added knowledge of the structure of fundamental objects such as viruses or stars is of paramount importance in their respective sciences. From the real-world standpoint, physical reality uses both discrete structures (in quantum mechanics) and the continuum (elsewhere). We have not yet discerned the ultimate nature of reality—possibly a greater knowledge of the continuum would enable us to make strides in that direction.

Additionally, computations made using the assumptions of the continuum are often much simpler. If the continuum is abandoned, there are no circles—just a bunch of disconnected dots equidistant from the center. One would not walk around a circular pond, traversing a distance of two times pi times the radius of the pond, but would walk in a sequence of straight line segments from dot to adjacent dot. The computation of the length of such a path would be arduous—and would turn out to equal $2\pi r$ to an impressive number of decimal places. The circle is a continuum idealization that does not exist in the real world—but the practical value of the circle and the simplifying computations it entails are far too valuable to be summarily abandoned.

Finally, the quest for models that satisfy different systems of axioms often has surprising consequences for our understanding of the real world. Attempts to derive models in which Euclid's parallel postulate was not satisfied led to the development of hyperbolic geometry, which was incorporated in Einstein's theory of relativity, the most accurate theory we have on the large-scale structure and behavior of the Universe. As Nikolai Ivanovich Lobachevsky put it, "There is no branch of mathematics, however abstract, which may not some day be applied to phenomena of the real world."[11]

NOTES

1. This quote is from Plato's *Theaetetus*, section 152a. More on Protagoras can be found at http://en.wikipedia.org/wiki/Protagoras. Even though Wikipedia is user-edited, my experience has been that it's accurate when dealing with mathematics, physics, and the histories thereof—possibly because no one has any dog in the race, possibly because there isn't even a dog race with regard to matters such as this.

2. This quote is so famous that most sources just reference Einstein! The vast majority of its occurrences seem to be from math teachers who, like myself, wish to put their students at ease. Many people think that Einstein was a mathematician rather than a physicist, but his only mathematical contribution of which I am aware is the "Einstein summation convention," which is essentially a notation—like inventing the plus sign to denote addition.

3. Even the Securities and Exchange Commission warns against them. See http://www.sec.gov/answers/ponzi.htm.

4. Carl B. Boyer, *A History of Mathematics* (New York: John Wiley & Sons, 1991), p. 570.

5. Ibid.

6. Ibid.

7. See http://archives.cnn.com/2002/WORLD/europe/04/24/uk.kissinger/.

8. L. Wapner, *The Pea and the Sun (A Mathematical Paradox)* (Wellesley, Mass: A. K. Peters, 2005). This is a really thorough and readable exposition of all aspects of the Banach-Tarski theorem—including an understandable treatment of the proof—but you'll still have to be willing to put in the work. Even if you're not, there's still a lot to like.

9. See http://mathworld.wolfram.com/Zermelo-FraenkelAxioms.html. You'll have to fight your way through standard set theory notation (which is explained at the top of the page) in order to understand them, but the axioms themselves are pretty basic. There is a link and a further explanation for each axiom. Most mathematicians never really worry about these axioms, as the set theory they use seems pretty obvious, and are only concerned with finding a useful version of the axiom of choice (there are others besides the well-ordering principle). The two industry standard versions that I have found most useful are Zorn's Lemma and transfinite induction, and I believe that's true for the majority of mathematicians.

10. H. Weyl, *The Continuum* (New York: Dover, 1994), p. xii. Hermann Weyl was one of the great intellects of the early portion of the twentieth century. He received his doctorate from Göttingen; his thesis adviser was David Hilbert. Weyl was an early proponent of Einstein's theory of relativity, and studied the application of group theory to quantum mechanics.

11. Quoted in N. Rose, *Mathematical Maxims and Minims*, Raleigh N.C.,: Rome Press, 1988).

Reality Checks

Pascal's Wager

The French mathematician and philosopher Blaise Pascal was probably the first to combine philosophy and probability. Pascal was willing to acknowledge the possibility that God might not exist, but argued that the rational individual should believe in God. His argument was based on the probabilistic concept of expectation, which is the long-term average value of a bet. If you bet that God existed and won, the payoff was life ever after—and even if the probability that God existed was small, the average payoff from making this bet dwarfed the average payoff you would receive if God did not exist. A slightly different version of this is to look under the streetlight if you lose your car keys one night—the probability of the keys being there may be small, but you'll never find them where it's dark.

As the nineteenth century dawned, some of the leading thinkers of the era noted the success of physics and chemistry, and tried to apply some of the ideas and results to the social sciences. One such individual was Auguste Comte, who was one of the creators of the discipline of sociology, which is the study of human social behavior. His treatise, *Plan of Scientific Studies Necessary for the Reorganization of Society*, outlined his philosophy of positivism. Part of this philosophy can be expressed in terms of the relation between theory and observation—as Comte put it, "If it is true that every theory must be based upon observed facts, it is equally true that facts can not be observed without the guidance of some theory. Without such guidance, our facts would be desultory and fruitless; we could not retain them: for the most part we could not even perceive them."[1]

Simon Newcomb made significant contributions to astronomy and mathematics. He was a computer—in the days when this described a job occupation rather than an electronic device—and oversaw a program that revised the calculation of the positions of the astronomical bodies. He helped Albert Michelson calculate the speed of light, and also helped refine the calculation of the amusingly named Chandler wobble, the change of spin of Earth around its axis. Newcomb did not confine himself to the physical sciences; his *Principles of Political Economy* (1885) was praised by the famed economist John Maynard Keynes as "one of those original works which a fresh scientific mind, not perverted by having read too much of the orthodox stuff, is able to produce from time to time in a half-formed subject like economics."[2] High praise, indeed, from one of the leading economists of the twentieth century. To cap a distinguished career, Newcomb was buried in Arlington National Cemetery, and President Taft attended the funeral.

Obviously, both these individuals were among the leading intellectuals of their times—but they are both known for making predictions that would make the all-time Top 100 list under the heading of "Predictions You Wish You Hadn't Made—At Least, Quite So Publicly." Comte wrote a philosophical treatise examining things that would never be known, including in his list the chemical composition of the stars. Several years later, Robert Bunsen and Gustav Kirchhoff discovered spectroscopy, and the analysis of the spectrum of light emitted by stars permitted their chemical composition to be deduced. Newcomb was interested in powered flight, but did calculations—later shown to be erroneous—that convinced him that such was impossible without the development of new methods of propulsion and much stronger materials. A few years later, Orville and Wilbur Wright achieved powered flight with not much more than a wooden frame, wires for control, and an internal-combustion engine.

As Niels Bohr so wryly observed, "Prediction is difficult—especially of the future."[3] Predicting what can or cannot be known in the area of mathematics is also difficult, but since most such predictions involve fairly arcane areas of study, they usually do not register on the public's radar screen. However, predictions regarding the limitations of knowledge and achievement in the physical world are much more likely to come under scrutiny—and when one predicts that we will *never* know the chemical composition of the stars, it takes an extremely long time to be proved correct. Making such predictions would seem to be a losing intellectual proposition—like taking the wrong side of Pascal's Wager. You can always be proven wrong, and you are extremely unlikely to be proven right.

It's Tough to Be a Physicist

One cannot help but be impressed by the extraordinary success of physics, a success to which mathematics makes a substantial contribution. I remember being amazed as a child when the *New York Times* published details of a partial solar eclipse that was to occur that day. The article included the time of onset, the time of maximum coverage, the time of conclusion, and a graphic of the path of the eclipse—in which portions of the country one would be able to view this phenomenon. To think that a few laws propounded by Isaac Newton, coupled with some mathematical calculations, enable one to predict such phenomena with almost pinpoint accuracy is still a source of substantial wonder, and unquestionably represents one of the great triumphs of the human intellect.

Most of the great theories of physics represent the scientific method in full flower. Experiments are conducted, data is gathered, and a mathematical framework explaining the data is constructed. Predictions are made—if those predictions involve as-yet-unobserved phenomena whose existence is later validated, the theory attains greater validity. The discovery of the planet Neptune gave added weight to Newton's theory of gravitation, the precession of the perihelion of Mercury helped substantiate Einstein's theory of relativity.

Physics is sometimes thought of as being simply a branch of applied mathematics. I feel this does a severe injustice to physics. The difference between physics and mathematics is somewhat akin to the difference between the art of painting portraits and abstract expressionism. If you are hired to paint a portrait, it has to end up looking like the person whose portrait is being painted. Insofar as my limited understanding of abstract expressionism goes, anything you feel like putting on canvas qualifies as abstract expressionism—at least if it's so abstract that nobody can recognize what it is. This is a bit unfair to mathematics, some of which is highly practical—but some of it is so esoteric as to be incomprehensible to anyone but a specialist, and totally useless for any practical purpose. My appreciation for abstract expressionism, as well as my understanding of it, is limited—but I might look at it anew, considering that a Jackson Pollock recently sold for $140 million. Maybe this analogy is not so bad, because highly abstract areas of mathematics have turned out to have significant—and unexpected—practical value, and $140 million is a lot of practical value.

The successes of physics are extraordinary—but its failures are extraordinary, too.

One of the early theories of heat was the phlogiston theory. The phlogiston theory states that all flammable substances contain phlogiston, a colorless, odorless, weightless substance that is liberated in burning. I strongly doubt that anyone ever produced a truly axiomatic theory of phlogiston, but if anyone did, the moment that Antoine Lavoisier showed that combustion required oxygen, the phlogiston theory was dead as the proverbial doornail. No further treatises would be written on phlogiston theory because it had failed the acid test—it did not accord with observable reality. This is the inevitable fate that awaits the beautiful physical theory that collides with an ugly and contradictory fact. The best that can be hoped for such a theory is that a new one supersedes it, and the old theory is still valid in certain situations. Some venerable theories are so useful that, even when supplanted, they still have significant value. Such is the case of Newton's law of gravitation, which still does an admirable job of predicting the vast majority of everyday occurrences, such as high and low tides on Earth. Even though it has been superseded by Einstein's theory of general relativity, it is fortunately not necessary to use the tools of general relativity to predict high and low tides, as those tools are considerably more difficult to use.

Mathematics rarely worries about reality checks. There are exceptions, such as the tale related to me by George Seligman, one of my algebra instructors in college, whose classes I greatly enjoyed. The real numbers—the continuum discussed in the previous chapter—form a certain type of algebraic system of dimension 1.[4] The less-familiar complex numbers (built up from the imaginary number $i = \sqrt{-1}$) form a similar structure of dimension 2, the quaternions one of dimension 4, and the Cayley numbers a structure of dimension 8. Seligman said that he had spent a couple of years deriving results concerning the structure of dimension 16 and was ready to publish them when someone showed that no such structure existed, and that the four known structures described above were all there were. Interestingly, at that time two manuscripts had been submitted for publication to the prestigious *Annals of Mathematics*. One of the papers outlined the structure of the object of dimension 16; the other showed that no such object existed. For Seligman, it was two years of work down the drain, but despite that setback he had a long and productive career.

For the most part, though, mathematics is extremely resilient with regard to the question of how many angels can dance on the head of a pin. While the question is open, mathematicians can write papers in which they derive the consequences of having a particular number of dancing angels, or of placing upper or lower limits on the number of angels. If the question is eventually answered, even the erroneous results can be viewed as steps leading toward the solution. Even if it is shown that this is a question that cannot be answered, a perfectly reasonable approach is to add an axiom regarding the existence or nonexistence of dancing angels and to investigate the two systems that result—after all, this was the approach that was followed when it was shown that the continuum hypothesis was independent of the axioms of Zermelo-Fraenkel set theory. The physicist, ever mindful that his or her results must accord with reality, is indeed like the portrait photographer; whereas the mathematician, like the abstract expressionist, can throw any array of blobs of paint on a canvas and proudly proclaim that it is art—as did the English mathematician G. H. Hardy, whom we encountered in the introduction.

The Difference Between Mathematical and Physical Theories

The word *theory* means different things in physics and mathematics. The dictionary does a good job of explicating this difference—a theory in science is described as a coherent group of general propositions used as explanation for a class of phenomena, whereas a theory in mathematics is a body of principles, theorems, or the like belonging to one subject. My library contains books on the theory of electromagnetism and the theory of groups. Even though the theory of groups is not my area of expertise, I have little difficulty navigating my way through it. On the other hand, I got a D in electromagnetism in college (to be fair, that was the first semester that I ever had a girlfriend, and so my attention to the course in electromagnetism was unquestionably diverted), and one of my goals on retirement is to read the book through to its conclusion. In my spare moments, I have picked it up and started reading—it's still *really* tough sledding. It's not the mathematics that's the problem—it's the juxtaposition of mathematics and an understanding of, or a feel for, physical phenomena.

A mathematical theory generally starts with a description of the objects under investigation. Euclidean geometry is a good example. It starts with the following axioms, or postulates.

1. Any two points can be joined by a straight line.
2. Any straight line segment can be extended indefinitely in a straight line.
3. Given any straight line segment, a circle can be drawn having the segment as radius and one endpoint as the center.
4. All right angles are congruent.
5. Through any point not on a given straight line, one and only one straight line can be drawn parallel to the given line through the given point.

Certain nouns are not defined (point, straight line, etc.), and neither are certain verbs (joined, extended indefinitely, etc.), although we all know what they mean. Once we accept these axioms, *in the sense that we agree to work with them,* the game is afoot—derive logical conclusions from them. That's all the mathematician has to do.

The theory of electromagnetism starts with Coulomb's law, which states that the magnitude of the electrostatic force between two point charges is directly proportional to the magnitudes of each charge and inversely proportional to the square of the distance between the charges. This law is analogous to Newton's law of universal gravitation, which states that the magnitude of the gravitational force between two point masses is directly proportional to the mass of each object and inversely proportional to the square of the distance. The reason the two theories are not identical is that mass is inherently positive, whereas charge can be either positive or negative. We accept Coulomb's law as the starting point because no measurement has ever contradicted it. The game is again to derive logical conclusions from it—but that is far from all that the physicist has to do. The logical conclusions enable the physicist to devise experiments that will test not the validity of the conclusions—which is all that matters in mathematics—but whether the conclusion is consistent with observable reality. Logical conclusions in physics are continually subjected to this reality check—because the utility of a physical theory is limited by how it accords with observable reality.

When Two Theories Overlap

Physicists have developed two highly successful theories: relativity, which does an excellent job of describing the gravitational force, and quantum mechanics, which does an even better job (at least, from the standpoint of the precision to which experimentation has confirmed the two theories) of describing the mechanical and electromagnetic behavior of particles at the atomic and subatomic level. The problem is that relativity only manifests itself in the realm of large objects, whereas quantum mechanical effects are significant only in the world of the really, really, *really* small. Many physicists agree that the single most important theoretical problem confronting physics is the construction of a theory (generally referred to as quantum gravity) that subsumes both these theories. Current contenders include string theory and loop quantum gravity,[5] and part of the difficulty in picking a winner is devising or discovering phenomenological results that will help distinguish between the two. After all, if the two theories predict different results when five black holes simultaneously merge, it might be a long wait for such an event to occur.

The melding of theories in mathematics is seamless by comparison. Probably the first to achieve success in this area was Descartes, who wrote an appendix to his *Discourse on Method* which laid the foundation for analytic geometry. In terms of utility, the few pages Descartes wrote on analytic geometry far outstrip the volumes he wrote on philosophy, as analytic geometry enables one to apply the precise computational tools of algebra to geometric problems. Ever since then, mathematicians have been happily co-opting results from one area and applying them to another. Topology[6] and algebra are, on the surface, disparate fields of study. However, there are important results in topology obtained by using algebraic tools such as homology groups and homotopy groups (the precise definition of a group will be given in chapter 5) to study and classify surfaces, and there are equally valuable results obtained by taking advantage of the topological characteristics of certain important algebraic structures to deduce algebraic properties of these structures.

Part of the charm of mathematics, at least to mathematicians, is how results in one area can often be fruitfully used in another apparently unrelated area. My own area of research in recent years was fixed-point theory. A good example of a fixed point is the eye of a hurricane; while all hell breaks loose around the hurricane, the eye experiences not even a gentle zephyr. Many fixed-point problems are nominally placed in the domain of real analysis, but at the same time that I and a colleague submitted a solution to a particular problem that placed heavy reliance on combinatorics, the branch of mathematics that deals with the number and type of arrangements of objects, a mathematician in Greece submitted a paper solving the same problem, also using combinatorics, but an entirely different branch of combinatorics than the one that I and my colleague employed. It hasn't happened yet, but I wouldn't be surprised to see a conference on combinatorial fixed-point theory at some stage in the future.

The Standard Model

When I studied physics in high school and college, atoms were portrayed as consisting of a nucleus of protons and neutrons, with electrons orbiting around the nucleus in a manner akin to planets orbiting a star (although some of my teachers did mention that this was not a totally accurate depiction). There were four forces—gravity, electromagnetism, the weak force (which governed radioactivity), and the strong force (which held nuclei together against the mutual repulsion of the positively charged protons in the nucleus). There were a few leftover particles, such as neutrinos and muons, and although it was understood that electromagnetism was the result of movement of electrons, the jury was still out on how the other forces worked.

Half a century later, much of this has been augmented and unified as the Standard Model.[7] It is now known that there are three families of particles that admit a very attractive classification scheme, and that forces are conveyed through the interchange of various particles. However, even if the Standard Model is the last word, there are still numerous questions, such as "What causes mass?" (the current leading contender is something called the Higgs particle, which no one has yet seen and which always seems to be one generation of particle accelerators away) and "Why is electromagnetism stronger than gravity by a factor of 1 followed by 39 zeroes?"

One of the attractive features of a theory of quantum gravity is that it should allow for unification of the four forces. Nearly thirty years ago, Sheldon Glashow, Steven Weinberg, and Abdus Salam won the Nobel Prize for a theory[8] that unified the electromagnetic force and the weak force into the electroweak force, which was present only at the ultrahigh temperatures that occurred immediately after the big bang. A number of physicists believe that there is a theory in which all the forces will coalesce into a single force at an almost inconceivably high temperature, and then the various individual forces will separate as the temperature falls, somewhat as various components of a mixture will separate from the mixture as it cools.

I'd love to see such a theory. I'm sure it would take me years of study before I had a hope in hell of comprehending it, for such a theory would undoubtedly be vastly different from any branch of mathematics I've ever studied. Most mathematical theories begin with a very general structure that has a relatively small set of axioms and definitions—such as the structure known as an algebra. A good example of an algebra is the collection of all polynomials—you can add and subtract polynomials and multiply them by constants or other polynomials, and the result is still a polynomial. Division, however, is not an allowed operation—just as some integers divided by other integers are not integers (such as 5 divided by 3), some polynomials divided by other polynomials are not polynomials.

The study of algebras proceeds by adding other hypotheses. Algebras beget Banach algebras, which beget commutative Banach algebras, which beget commutative semi-simple Banach algebras—each additional adjective representing an additional hypothesis (or hypotheses). Physics doesn't seem to follow this scheme—the axioms of a theory are constantly subject to review. In fact, the Standard Model is not so much the deductions as it is *the* Model itself—the deductions available from the hypotheses are generally used not as a way to build a better refrigerator, but as a check upon the validity of the Model.

The Limitations of Physics

It is generally within the last century that physics has come to grips with its own limitations. Although the Standard Model talks about particles and forces, one of the more modern ideas in physics is that information is just as much of a fundamental concept. In particular, much of what we have discovered concerning the limitations of physics can be classified in terms of information.

Some of these limitations occur because the information we need is simply not accessible, if indeed it exists at all. We cannot know what happened before the big bang—if indeed anything did—because information travels no faster than the speed of light. Nor can we know what lies over the hill—if there is a portion of the universe that is farther away from us in light-years than the time since the big bang, and if that portion is receding from us faster than the speed of light, no information from this portion will ever reach us.

Some limitations are imposed because there is a limit to the accuracy of the information concerning it. Heisenberg's famous uncertainty principle tells us that the more accurately we are able to ascertain the position of a particle, the less accurately we can know its momentum (or, as is more commonly thought, its velocity). The consequences of the uncertainty principle and other aspects of quantum mechanics, which will comprise a significant portion of the next chapter, are among the most eye-opening and counterintuitive results in the history of human knowledge. This limitation also hampers our ability to predict—negating Laplace's famous statement concerning omniscience. We could say that the universe prevents us from knowing how things will be by concealing from us how things are.

When Theories Do Battle

In the middle of the twentieth century, there were two main contenders to explain the fact that over large scales of space and time, the universe appeared unchanging. Although the big bang theory,[9] which posited the creation of the Universe in an enormous explosion, was to emerge triumphant, it had a strong rival in the steady state theory. One of the key assumptions of the steady state theory[10] was that one atom of hydrogen was created *de nihilis* per 10 billion years in every cubic meter of space. That's not a whole lot of creation—but it requires abandoning the principle of matter-energy conservation that is nominally one of the bedrock principles of physics. However, there are limits to which scientific principles can be confirmed by experiment—and in the 1950s (and possibly now as well), it was impossible to measure with a precision that would invalidate such a result.

There is an uncertainty (which has nothing to do with the uncertainty principle) that surrounds any set of hypotheses in physics. The best one can do with any set of hypotheses is to make deductions and test them by experiment, and the precision of all experiments is limited. In order to observe the creation of one atom of hydrogen per 10 billion years per cubic meter, one can't just pick a cubic meter and observe it for 10 billion years. Even granted the fact that it would be hard to find someone or something willing to sit and watch a cubic meter of space for that long a time, you might be unfortunate and pick a cubic meter in which nothing happens—the steady state theory obviously talks about an average rather than an exact occurrence. The steady state theory did not fall by the wayside because atom creation went unobserved—it fell because in an unchanging universe, there would have been no cosmic microwave background. Such a background was predicted by the big bang theory to be a relic of the big bang—and when it was observed by Arno Penzias and Robert Wilson in the 1960s, the big bang theory emerged as the clear-cut winner.

Physics is frequently confronted with situations in which it must rely on statistical methods rather than observations—various theories that have predicted that protons decay, but with exceedingly long time intervals before they do, so the solution is to assume that there is a frequency distribution with which those protons decay and watch a large number of protons. Many physical theories are confirmed or refuted on the basis of statistical tests—not unlike theories in the social sciences, except for the fact that theories in the social science are often accepted or rejected on the basis of confirmation at the 95 percent level of confidence, whereas physical theories must meet much more stringent criteria.

Theories in mathematics never do battle in this fashion, and they are never resolved on the basis of statistical evidence. As with a great problem such as the truth or falsehood of the continuum hypothesis, the resolution of the problem adds something new to mathematics. It is true that theories fall in and out of favor with the mathematical community, and it is also true that theories are sometimes supplanted by more all-encompassing theories. If there are competing explanations for phenomena in the real world, mathematics may provide some of the tools needed to resolve the dispute, but without experiments and measurement these tools are essentially useless.

The last chapter in this section concerns which mathematical model best describes the small-scale structure of our universe—discrete structures or the continuum. Both theories, from a mathematical standpoint, are equally valid—but when it came to description of the universe, there could be only one winner.

NOTES

1. See http://en.wikipedia.org/wiki/Auguste_Comte. As I have mentioned, Wikipedia biographies are generally reliable, and often very well documented.
2. See http://en.wikipedia.org/wiki/Simon_Newcomb.
3. See http://sciencepolicy.colorado.edu/zine/archives/31/editorial.html. A quick Google search also finds that this quote is attributed to Mark Twain, who said lots of very clever things, and Yogi Berra, who said lots of things like this, and as a result gets a lot of credit for things like this which he may, or may not, have said.
4. According to Seligman, the precise problem was to determine for which values of n there exists a bilinear map (the multiplication) of $\mathbb{R}^n \times \mathbb{R}^n \mapsto \mathbb{R}^n$ such that $ab = 0$ if and only if either $a = 0$ or $b = 0$. If you're not familiar with the notation, $\mathbb{R}^n$ is the set of all n-dimensional vectors whose components are real numbers. Bilinear maps are generalizations of the distributive law in both variables—$(a + b)c = ac + bc$ and $a(b + c) = ab + ac$. Additionally, because a and b are vectors, a bilinear map must satisfy $(ra)b = r(ab)$ and $a(rb) = r(ab)$ for any real number r.
5. This is a terrific opportunity to plug two immensely enjoyable best sellers by Brian Greene, *The Elegant Universe* (W. W. Norton 1999) and *The Fabric of the Cosmos* (Alfred A. Knopf, 2004). Despite what reviewers may say, these wonderful books are tough sledding—deep ideas never admit easy explanations, and both string theory and loop quantum gravity are incredibly deep ideas. Nonetheless, Greene does an excellent job with string theory in the first book, but since he is a believer in string theory, loop quantum gravity is given relatively short shrift. To be fair, loop quantum gravity is unquestionably a minority position in the physics community—but the right of a minority to become a majority is nowhere more religiously observed than in physics.
6. Topology is the study of the properties of geometric figures or solids that are not changed by deformations such as stretching or bending. The classic example is

that a doughnut is topologically equivalent to a coffee mug, because each has precisely one hole (you know where it is in the doughnut; the hole in the coffee mug is where you put your finger through when you hold the mug). If you had a piece of clay, and poked a hole in it, you could shape it like a doughnut (easy) or a coffee mug (not so easy) by stretching and bending the clay without any further tearing.

7. See http://en.wikipedia.org/wiki/Standard_Model. This is an excellent short exposition of the Standard Model, along with a beautiful chart that puts the periodic table to shame. You have to click several times on the chart before you get to a readable resolution, but it's worth it.

8. See http://en.wikipedia.org/wiki/Electroweak. The first two paragraphs give you all you need, but if you like staring at equations, there's a nice little window that has the basic equations of the theory—if $E = mc^2$ is the most impressive equation you've ever seen, take a look. Because Wikipedia is user composed, the depth of treatment in different sections varies wildly. I'm not a physicist, but I can recognize the symbols and what the equations are saying, but I have no idea where they come from and how they might be used.

9. See http://en.wikipedia.org/wiki/Big_Bang. If explanatory Web sites were rated on a 1 to 10 scale, this one would be a 10—it's as good as it gets. Good graphics, understandable explanations, excellent hyperlinking—this site is so good if it had pop-up ads, you wouldn't mind it.

10. See http://en.wikipedia.org/wiki/Steady_State_theory. This site is nowhere near as impressive as the one for the big bang theory. No graphics, a rather perfunctory explanation, but that's not really surprising, because the steady state theory is dead, dead, dead. I imagine that sighs of relief were heard throughout the astrophysical community when this theory bit the dust, because matter-energy conservation is so fundamental a principle you'd hate to abandon it.

3 All Things Great and Small

Glamour vs. Meat and Potatoes

The theory of relativity is probably the most glamorous result of twentieth-century physics. It is both beautiful and profound, and it made Albert Einstein an iconic figure. However, other than to show the equivalence of matter and energy, which has led to extremely destructive weapons and an energy technology which is widely used outside the United States but has fallen into disrepute here, exactly what has the theory of relativity done for the average person?

The short answer is "not much." The theory of relativity also involves gravity, but although gravity is used to turn dynamos much as gravity was used to turn waterwheels in the past, it is the electricity produced by the dynamos that powers our lives, not the gravity-induced fall of the water that turns them. Unquestionably, the theory of relativity has had a significant impact upon the world, but it pales in comparison with the impact made by the study of the physics of the electron and the photon.

The deeper understanding of the electron and the photon is the domain of quantum mechanics. Many great physicists have contributed to quantum mechanics, including Einstein, but there was no previous Isaac Newton in this branch of physics to be knocked off his or her pedestal. Yet quantum mechanics has changed our lives, perhaps more than any single branch of physics has ever done—although the classical theory of electromagnetism would be a strong contender. But quantum mechanics has been much more than a generator of technology; it has greatly changed and challenged our understanding of the nature of reality.

What Does It All Mean?

Ever since Pythagoras proved what is arguably the most important theorem in mathematics, mathematics has generally had a very clear view of what it is trying to accomplish. Pythagoras knew, as it had been known since Egyptian times, that some of the classic triangles were right triangles, such as the triangle with sides 3, 4, and 5. Noticing that $3^2 + 4^2 = 5^2$, he was able to generalize this to show that in a right triangle, the square of the hypotenuse was equal to the sum of the squares of the other two sides. He knew what he wanted to prove, and when he proved it he knew what he had—a theorem so important that he ordered a hundred oxen to be barbecued in celebration. I sometimes tell my students this tale, adding that this provides a measuring rod for the importance of mathematical theorems. The fundamental theorem of arithmetic (that every number can be uniquely expressed as a product of primes), the fundamental theorem of algebra (that every nth-degree polynomial with real coefficients has n complex roots), and the fundamental theorem of calculus (that integrals can be evaluated via anti-differentiation) are all sixty-oxen theorems, and to my mind, no other theorems come close to those.

It's different in physics—especially in quantum mechanics. Both physicists and mathematicians "play" with what they have in an attempt to deduce new and interesting results, but when mathematicians deduce such a result, they almost never have to worry about what it means. It is what it is, and the next step is to find applications of the result, or deduce new consequences from it.

Physicists, on the other hand, have to decide what the result means—what the mathematics actually represents in the real world. Quantum mechanics is such an incredibly rich and profound area that physicists are debating the meaning of results nearly a century old. Niels Bohr, one of the architects of the theory, expressed this sentiment perfectly when he declared, "If quantum mechanics hasn't profoundly shocked you, you haven't understood it yet."[1]

Richard Arens

My first teaching job took me to UCLA in the fall of 1967, a few years after the release of the film *Mary Poppins*. One of the supporting actors in the cast was the venerable British comedian Ed Wynn, who played the role of Mary Poppins's Uncle Albert. At the time I arrived at UCLA, one of the senior members of the mathematics department was Richard Arens, who bore a striking physical resemblance to Ed Wynn—he had a bald head with a fringe of hair surrounding it, and an air of perpetual amusement.

In the course of my work, I had occasion to read several papers that Arens had written. These papers were a treat—they contained interesting and unexpected results, almost invariably proved in an interesting and unexpected way (many results in mathematics are proved by techniques so well known that a few lines into the proof you can say to yourself something like "Cantor diagonal proof"—this was used to show that the set of all infinitely long names cannot be matched one-to-one with the positive integers—and skip to the next section).

At one stage in his impressive career, Arens decided that what was needed was for a mathematician to look at quantum mechanics. He did so for a number of years. I talked to him about it, and he said that he had studied it intensively, and had basically gotten nowhere. I suspect that "nowhere" for Richard Arens was a lot further than it might be for others, but nonetheless it indicates the depth and complexity that appears in quantum mechanics.

Any Questions?

For a number of years, I was the graduate adviser in the Mathematics Department at CSULB. One of my jobs was to keep tabs on our teaching associates, the graduate students we supported by having them teach lower-level classes. At the start of each year, I gave a short talk on what I considered generally good advice for teaching. One of the issues concerned how to handle perplexing questions. I told them that every so often a student would ask them a question that they couldn't answer off the top of their head. It's happened to me, and I'm sure to practically every other math teacher as well. I told them that in that case, they should say, "That's a very interesting question. Let me think about it and get back to you on it." In doing this, they have paid respect, both to the questioner and the question, and have kept faith with what is one of the essential missions of a teacher—to answer questions as well as possible. Sometimes the correct answer to a question requires work, and it is more important to have the correct answer later than an incorrect answer now.

I'd like to give the same advice to readers of this book—especially in this chapter—but sometimes the answers are not known, even to the best minds in physics, and certainly not to me. So I would ask for a certain measure of indulgence on the part of the reader. What quantum mechanics has shown us about the nature of reality and the limitations of knowledge is truly fascinating—but the final version of this saga is a long way from being written, and may never be written. Unquestionably, though, what we have learned via quantum mechanics about reality and the limitations of knowledge is so fascinating and compelling that this book would be incomplete without a discussion of this subject.

Max Planck and the Quantum Hypothesis

As the nineteenth century came to a close, physicists around the world were beginning to feel their time had come and gone. One physicist advised his students to pursue other careers, feeling that the future of physics would consist of the rather mundane task of measuring the physical constants of the universe (such as the speed of light) to ever-increasing levels of accuracy.

Still, there were (apparently) minor problems that had yet to be resolved. One of the unsettled questions concerned how an object radiates. When iron is heated on a forge, it first glows a dull red, then a brighter red, and then white; in other words, the color changes in a consistent way with increasing temperature. Classical physics was having a hard time accounting for this. In fact, the prevailing Rayleigh-Jeans theory predicted that an idealized object called a blackbody would emit infinite energy as the wavelength falling on it became shorter and shorter. Short-wavelength light is ultraviolet; the failure of the Rayleigh-Jeans theory to predict finite energy for a radiating blackbody exposed to ultraviolet light came to be known as the "ultraviolet catastrophe."[2]

When a scientific theory encounters an obstacle, several different things can happen. The theory can overcome the obstacle; this frequently occurs when broader ramifications of a new theory are discovered. The theory can undergo minor revisions; like software, the alpha version of a theory is often in need of fine-tuning. Finally, since any scientific theory is capable of explaining only a limited number of phenomena, it may be necessary to come up with a new theory.

The Rayleigh-Jeans theory operated under a very commonsense premise—that energy could be radiated at all frequencies. An analogy would be to consider the speed of a car—it should be able to travel at all velocities up to its theoretical limit. If a car cannot go faster than 100 miles per hour, for instance, it should be able to move at 30 miles per hour, or 40 miles per hour, or 56.4281 miles per hour. However, writing down a few numbers is somewhat deceptive, because they are all rational numbers. As we know from the previous chapter, there are uncountably many real numbers less than 100.

One day in 1900, the German physicist Max Planck made a bizarre assumption in an attempt to escape the ultraviolet catastrophe. Instead of assuming that energy could be radiated at all frequencies, he assumed that only a finite number of frequencies were possible, and these were all multiples of some minimum frequency. Continuing the analogy with the speed of the car, Planck's hypothesis would be that something like only speeds that were multiples of 5—25 miles per hour, 40 miles per hour, etc.—would be possible. He was able to show almost immediately that this counterintuitive hypothesis resolved the dilemma, and the radiation curves he obtained from making this assumption matched the ones recorded by experiment. That day, while walking with his young son after lunch, he said, "I have had a conception today as revolutionary and as great as the kind of thought that Newton had."[3]

His colleagues did not immediately think so. Planck was a respected physicist, but the idea of the quantum—energy existing only at certain levels—was at first not taken seriously. It was viewed as a kind of mathematical trickery that resolved the ultraviolet catastrophe, but did so by using rules that the real world did not obey. Ever since Isaac Newton incorporated mathematics as an essential part of a description of natural phenomena, it has generally been easier for a theoretician to sit down with pencil and paper, and derive mathematical consequences, than it has for an experimenter to devise and carry out a successful experiment. As a result, there is sometimes the feeling that mathematics is merely a convenient language to describe phenomena, but it does not give us an intuitive insight into the nature of the phenomena.

Planck's idea languished for five years, until Einstein used it in 1905 to explain the photoelectric effect. Eight years later, Niels Bohr used it to explain the spectrum of the hydrogen atom. Within another twenty years, Planck had won a Nobel Prize, and quantum mechanics had become one of the fundamental theories of physics, explaining the behavior of the world of the atom and making possible many of the high-tech industries of today.

With the coming of the Nazis, German science suffered severely. Many of the leading scientists were either Jewish or had Jewish relatives, and fled the country. Many others reacted with abhorrence to the Nazi regime, and also departed. Planck, although deploring the Nazis, decided to stay in Germany. It was to be a tragic decision. In 1945, Planck's younger son was executed for his part in the "Revolt of the Colonels," the unsuccessful attempt by several members of the German armed forces to assassinate Hitler.

The Quantum Revolution Continues

Max Planck's revolutionary idea did more than simply resolve the ultraviolet catastrophe. Possibly only one other moment in science has opened the door to such an unexpected world—when Anton von Leeuwenhoek took his primitive microscope and examined a drop of water, only to discover forms of life never suspected and never before seen.

The quantum revolution has changed our world—technologically, scientifically, and philosophically. Much of the incredible technology that has been developed since the 1930s—the computer, the medical scanners, lasers, everything with a chip in it—are the result of the application of quantum theory to understanding the behavior of the subatomic world. Quantum mechanics has not only spawned sciences that did not exist prior to its conception, it has also greatly enriched some of the more venerable areas of study, such as chemistry and physics. Finally, quantum mechanics has fostered discoveries so profound that they make us wonder about the essential nature of reality, a topic that has been a matter of fierce philosophical debate for millennia.

Libraries could be built housing only books devoted to discussions of quantum mechanics, so I'll just concern myself with three of the most perplexing topics in quantum mechanics: wave-particle duality, the uncertainty principle, and entanglement.

Is Light a Wave or a Particle?

Probably no question in science has created more controversy over a longer period of time than the nature of light. Greek and medieval philosophers alike puzzled over it, alternating between theories that light was a substance and that light was a wave, a vibration in a surrounding medium. Almost two millennia later, Isaac Newton entered the debate. Newton, when he wasn't busy with mathematics, mechanics, or gravitation, found time to invent the science of optics. As had others, Newton puzzled over the nature of light, but finally voted for the theory that light was a substance.

We know the characteristics of substances, but what are some of the characteristics of waves? Not all waves behave similarly. Sound, one classic example of a wave, can go around corners. Light doesn't. Water waves, another obvious type of wave, can interfere with each other. When two water waves collide, the resulting wave can be either stronger or weaker than the original waves—stronger where the high points of both waves reinforce each other, and weaker where the high points of one wave coincide with the low points of the other wave.

Such was the almost universal reverence in which Newton was held that few efforts were made to either validate or dispel the wave theory of light for more than a century, even though the noted physicist Christian Huygens (1629–1695) strongly favored the view that light was a wave phenomenon. The individual who would finally perform the definitive experiment was Thomas Young, a child prodigy who could read by age two and who could speak twelve languages by the time he was an adult. In addition to being a child prodigy, fortune had favored Young in other respects, as he was born into a well-to-do family.

Thomas Young was a polymath whose accomplishments extended into many of the realms of science, and even beyond. He made significant contributions to the theory of materials; Young's modulus is still one of the fundamental parameters used to describe the elasticity of a substance. Young was also an Egyptologist of note, and was the first individual to make progress toward deciphering Egyptian hieroglyphics.

After a brilliant performance as a student at Cambridge, Young decided to study medicine. Young was extremely interested in diseases and conditions of the eye. He constructed a theory of color vision, observing that in order to be able to see all colors, it was only necessary to be able to see red, green, and blue. While still a medical student, he discovered how the shape of the eye changes as it focuses. Shortly after, he correctly diagnosed the cause of astigmatism, a visual fuzziness caused by irregularities in the curvature of the cornea.

Young's fascination with the eye led him to begin investigations into color vision and the nature of light. In 1802, he performed the experiment that was to show once and for all that light was a wave phenomenon.

The Double-Slit Experiment

Particles and waves behave differently as they go through slits. If one imagines waves coming onto shore, blocked by a jetty of rocks with one narrow opening, the waves spread out in concentric circles around the opening. If there are two narrow openings reasonably close to each other, the waves spread out in concentric circles around each opening, but the waves from each opening interact (the technical term is "interfere") with the waves from the other opening. Where the crests (where the waves are highest) of one set of waves encounter the crests from another set of waves, the cresting is reinforced. Where the crests from one set of waves encounter the troughs (where the waves are lowest) from the other set, they tend to neutralize each other, diminishing the amplitude of the crests where crests meet troughs.

The behavior of particles, on encountering a similar collection of narrow openings, is different. If two rectangular pieces of cardboard are lined up parallel to each other, a single narrow slit cut in the nearer of the two, and a paint sprayer directed at the nearer, a single blob of paint appears on the farther piece of cardboard directly behind the slit. The edges of the blob are not clearly defined, however, as paint particles spread out from the center but lessen in density the farther one is from the center. Cut two parallel slits in the nearer piece of cardboard and direct the paint sprayer at both, and similar blobs will appear on the farther piece of cardboard directly behind the slits.

Young constructed an experiment that took advantage of this difference. He cut two parallel slits into a piece of cardboard and shone a light through the slits onto a darkened background. He observed the alternating bright bands of light interspersed with totally dark regions. This is the classic signature of wave interference. The bright bands occurred where the "high points" (the crests) of the light waves coincided, the dark regions where the crests of one light wave were canceled out by the troughs of the other.

Einstein and the Photoelectric Effect

Young's double-slit experiment seemed to settle the issue of whether light was a wave or a particle—until Einstein put in his two cents' worth during his "miracle year" of 1905. One of the papers he wrote during this year explained the photoelectric effect. When light falls upon a photoelectric material, such as selenium, the energy in the light is sometimes sufficient to knock electrons out of the surface of the metal. Light produces electricity, hence the term *photoelectric*.

The wave theory of light predicted that the greater the intensity of the light, the greater should be the energy of the emitted electrons. In a classic experiment in 1902, Philipp Lenard showed that this was not the case, and that the energy of the emitted electrons was independent of the intensity of the light. No matter how strong the light source, the emitted electrons had the same energy. Lenard also showed that the energy of the emitted electrons depended upon the color of the incident light; if one used light of shorter wavelengths, the energy of the emitted electrons was higher than if one used light of longer wavelengths. This result also provides evidence how one's adviser, and the interests of that adviser, often influence the career of the student. Lenard's adviser at the University of Heidelberg was Robert Bunsen, who had discovered that the patterns of light, recognizable as bands of different color, characterized each element, and could be used to deduce the composition of the stars. This seminal experiment earned Lenard the Nobel Prize in 1905, the same year that Einstein was to explain the reasons behind the phenomena Lenard had discovered.

Einstein explained the photoelectric effect by invoking Planck's idea of quanta. He assumed that light behaved as a collection of particles (each particle is called a "photon"), with each photon carrying energy that depended upon the frequency of the light. The shorter the wavelength, the higher the energy of the associated photon. If you swing a bat faster, you will impart more energy to a baseball—assuming you hit it. When the short-wavelength (high-energy) photons hit an electron with enough energy to knock it out of the metal, that electron had more energy than one hit with a higher-wavelength (lower energy) photon—a Barry Bonds home run rather than a wind-assisted home run in Wrigley Field hit by a utility infielder.

The explanation of the photoelectric effect won Einstein the Nobel Prize in 1921. Great experiments such as Lenard's win Nobel Prizes, but great explanations such as Einstein's not only win Nobel Prizes but also make history. Perhaps unhappy at being upstaged by Einstein, possibly exacerbated by his inability to find the explanation for the photoelectric effect he had discovered (and on which he could have cornered both the experimental and theoretical markets), Lenard disparaged Einstein's theory of relativity as "Jewish science" and became an ardent supporter of the Nazis.

Is Matter a Wave or a Particle?

I have no idea how lengthy a typical doctoral dissertation is; I'm sure it varies with the field. Mine was about seventy typewritten pages and contained enough results that I was able to squeeze three published articles out of it—all of which have long been forgotten. I'm sure there are other, much lengthier doctoral dissertations, even in mathematics.

There are also shorter ones, *much* shorter. In 1924, Louis de Broglie wrote a very short dissertation in which he put forth the novel idea that matter could also have wavelike qualities. The core of his dissertation was a single equation expressing a simple relation between the particle's wavelength (obviously a wave property) and its momentum (a particle property). In 1927, this was experimentally confirmed, and de Broglie received the Nobel Prize in 1929.

To get a feel for this remarkable idea, imagine that we adjust the paint sprayer we described earlier so that the paint particles come out in a straight line, and very slowly—maybe one lonely particle of paint every few seconds. We aim this paint sprayer at the double-slit array, and after waiting for an agonizingly long period of time, look behind the slits to see what the rear piece of cardboard looks like. To no one's great surprise, it looks basically like it did when we turned the paint sprayer on full blast—two blobs with diffuse edges centered behind each of the two slits.

Perform this exact same experiment using, instead of a paint sprayer, an electron gun firing electrons instead of paint particles (and using a detector that records the impact of an electron by illuminating a pixel at the point of impact), and something weird and totally unexpected (well, except possibly by de Broglie) happens. Instead of seeing two blobs of light with diffuse edges, we see alternating dark patches and light patches—the signature of wave interference. The conclusion is inescapable—under these circumstances, the electron behaves like a wave. Matter, like light, sometimes behaves like a particle, sometimes like a wave.

Split Decisions—Experiments with Beam Splitters

A number of intriguing experiments in this area are conducted with beam splitters. Imagine that a photon starts its journey at home plate of a baseball diamond and hits a double, sliding into second base. In this experiment, however, the photon can get to second base via the usual route—going to first base and then to second—or by a path which in baseball would get the batter declared out—by going to third base and then to second. This is the modern version of the double-slit experiment. There is a light detector behind second base that records the impact of the photon, just as before; the paths that the photons can follow converge at second base so that wave interference, if it exists, can be detected. The beam splitter sends the photon by one of the two routes, via first or third base, and does so randomly but with equal probability of going via either route. In this variation, the light detector reveals interference patterns, as did the double-slit experiment; the photons are acting as waves.

Now change the experiment a little. Place a photon detector in the first-base coaching box (or the third-base coaching box, it doesn't matter). A coach can always tell when a runner has run past him—or whether no runner has gone by. Similarly, a photon detector can determine whether a photon has passed. This has a dramatic effect on the light pattern behind second base; it now consists of two bright patches, indicating that the photons have behaved as particles.

How Do Photons Know?

When observed (via a photon detector), photons behave as particles. When not observed (when there is no photon detector), photons behave as waves. This is strange enough—how does a photon know whether it is being observed or not? This is one of the riddles at the core of quantum mechanics, and it is a riddle that pops up in different guises.

Things get even stranger. In the 1970s, John Wheeler proposed a brilliant experiment, now known as a delayed-choice experiment. Position a photon detector far away from "home plate," and equip it with an on-off switch. If the photon detector is on, the photons behave as particles, if it is off, the photons behave as waves. This is essentially a combination of the two previous experiments.

Wheeler's suggestion was to turn the photon detector on or off *after* the photon has left home plate. This is known as a delayed-choice experiment, because the choice of whether to turn the detector on or off is delayed until after the photon has, presumably, already made *its* choice as to whether to behave as a wave or a particle. There appear to be two possibilities—the behavior of the photon is determined the instant it leaves home plate (but if so, how does it know whether the photon detector is on or off?), or the behavior of the photon is determined by the final state of photon detector. If the latter is the case (as experiments conclusively showed), the photon must simultaneously be in both states when it leaves home plate, or is in an ambiguous state that is resolved when either it passes the photon detector and learns it is being observed, or gets to second base without having been observed.

As has been previously mentioned, the mathematical description of quantum phenomena is done by means of probability. An electron, *before it is observed,* does not have a definite position in space; its location is defined by a probability wave, which gives the probability that the electron is located in a certain portion of space. Before it is observed, the electron is everywhere—although it is more likely to be in some places than others. Additionally, in going from here to there, it goes via all possible routes available in going from here to there! However, the observation process "collapses" the wave function, so that it can no longer be everywhere, and is instead somewhere in particular. The observation also collapses the electron's ability to go from here to there via all possible routes, and instead selects one route from the possible gazillions.

Wheeler also proposed that nature could illustrate how counterintuitive quantum mechanics is via a grandiose delayed-choice experiment. Instead of a beam splitter in a laboratory, a quasar billions of light-years away, acting as a gravitational lens, would do what the beam splitter does—allow the photon to come to Earth by one of two differing paths. These paths could be focused out in space; if no photon detectors had been placed along the paths, an interference pattern would result, and if there were photon detectors in place, the photons would act like particles. The counterintuitive aspect is that the photon, billions of years ago as it passed the gravitational lens, appears to have made the "decision" to behave as a wave or a particle. Experiments have shown that this decision is not made by the photon but by the universe—if an observation is made, the photon acts as a particle; if not, it acts as a wave.

Probability Waves and Observations: A Human Example

Arcane though the idea of probability waves and observations collapsing them may seem, there is a simple analogue that takes place annually at every university in the country. Many students enter as undeclared majors—not certain whether their futures lie in biochemistry, business, or something else. As a result, they take a diverse assortment of courses, encouraged by the university's general education policy of requiring students to take courses in a number of disciplines. These students are like probability waves; their as-yet-unselected majors are a probabilistic amalgam of biochemistry, business, and a whole bunch of other alternatives.

At some time, though, the student must select a major, usually done by conferring with an adviser who tells the student the options available, what the various majors require, and the career paths they allow (if the student does not already know), and the student makes his or her choice. This choice collapses the probability wave, and the student is now a declared major.

You're Nobody Till Somebody Observes You

A popular song from the 1950s was Dean Martin's "You're Nobody 'Till Somebody Loves You." In quantum mechanics, you're only a probability wave until someone, or something, observes you. What constitutes an observation in the physical universe, and when does it take place? A widely held view in the physics community is that an observation consists of an interaction with the universe. Our intuitive notion of reality—that things have definite states and attributes—collides with the world presented by quantum mechanics, in which things have a probabilistic mixture of states and attributes, and only interaction with the universe can create an actuality from what was initially just a possibility.

Schrödinger's Cat

Erwin Schrödinger came up with a tremendously provocative way to visualize the weirdness inherent in quantum behavior. He imagined a box containing a cat, a vial of poison gas, and a radioactive atom, which has a probability of 50 percent that it will decay within an hour. If it does, it triggers a mechanism releasing the poison gas, killing the cat (it seems likely that Schrödinger did not actually own a cat—although he may have owned one that was more trouble than it was worth). An hour goes by. In what state is the cat?[4]

The conventional answer to the question is that the cat is either dead or alive, and we'll know when we open the box. Quantum mechanics answers this question by saying it's half dead and half alive (or that it's neither)—and the answer will be determined when the box is opened, and observation collapses the wave function.

Counterintuitive though the half-dead, half-alive cat may be, that's the interpretation that quantum mechanics gives—and how can we refute it? Without an observation (which need not consist of actually looking at the cat, but simply obtaining information about the state of the radioactive atom whose decay determines the outcome), how can we know? Could your reclusive neighbor, whom you hardly ever see, be in a half-dead, half-alive state, which is only determined when he interacts in some way with the world? Just recently, a man was found in a mummified condition in front of a TV set—he had been dead (and the TV set on) for thirteen months before anyone decided to check up on him.

As a computational method, quantum mechanics is probably the most accurate in physics—confirmed to more decimal places than there are digits (including pennies) in the national debt. Some physicists feel that this is all physics can do—give computational rules that enable us to build computers and magnetic resonance imagers. A much larger number of physicists feel that this is telling us something deep and important about reality—but the physics community has not yet come to a consensus about what reality is, and if they can't, it's going to be difficult for the rest of us.

Quantum Erasers

The idea that photons and electrons are probability waves until they are observed, when they become objects, has been the subject of numerous experiments. One particular ingenious type of experiment is the quantum eraser experiment, first conceived by Marlon Scully and Kai Druhl in 2000. Going back to the baseball version of the setup, imagine that when a photon passes one of the coaches, that coach slaps it on the back (much like a baseball coach) with an identifying label that enables us to tell which route the photon takes. When that happens, an observation has clearly taken place, and the photon acts as a particle—the pattern on the detector behind second base is the familiar two blobs that characterize particles.

Now suppose that somehow, just as the labeled photons get to second base, the labels are removed (exotic as this labeling and unlabeling of photons might seem, there is a way to do it, but the details are not important for this discussion). There is then no evidence of the labeling—the labels have been erased (hence the term *quantum eraser*). With no evidence as to which way the photons got to second base, the interference pattern reemerges.

Bizarre—unquestionably. Surprising—no; this was precisely the result predicted by Scully and Druhl. We understand what quantum mechanics is telling us: that photons and electrons are probability waves until they interact with the universe, and then they are particles. If we cannot determine that they have interacted with the universe—and that's what quantum erasing accomplishes—they are probability waves. Among the things we may never know is why it is this way, and whether it could have been another way. This is one of the long-range goals of physics: to tell us not only the way the universe is, but why this is the only way it could be—or if it could be some other way.

It is a measure of how far we have come technologically that the May 2007 issue of *Scientific American* contains an article on how to build your own quantum eraser.[5] It doesn't seem very complicated—but whenever I try to assemble something, I always seem to have parts left over (why don't the manufacturers ever ship the right number of parts?). I remember reading an article on how, just prior to the test of the first atomic bomb, the physicists were concerned that the explosion might create an ultradense state of matter known as Lee-Wick matter, the appearance of which could (at least theoretically) result in the destruction of the universe. They convinced themselves that if that could happen, it would already have happened somewhere in the universe. However, they did not include my efforts to put things together in their calculations (I was only four years old at the time), and so I think I'll leave the construction of home quantum erasers to those with demonstrated mechanical ability.

The Uncertainty Principle

Some branches of mathematics, such as geometry, are highly visual; others, such as algebra, are highly symbolic, although many important results have been obtained by looking at algebraic problems geometrically or geometric problems algebraically. Nonetheless, most of us have a preference for looking at things one way or the other. Einstein had a beautiful way of expressing this: in his later years, he remarked that he hardly ever thought about physics by using words. Possibly, he saw pictures; possibly, he saw relationships between concepts. I marvel at this facility—while I sometimes think in terms of pictures, they are almost always derived from words describing them.

As physics probed ever deeper into the subatomic world in the first few decades of the twentieth century, it became harder and harder to visualize the phenomena that were occurring. As a result, some physicists, including Werner Heisenberg, preferred to treat the subatomic world through symbolic representation alone.

The Heisenberg who tackled this complex problem was very different from the Heisenberg who, at the end of World War I, was a "street-fighting man," engaging in pitched battles with Communists in the streets of Munich after the collapse of the German government following the war. Heisenberg was only a teenager at the time, and after the rebellious phase subsided, he switched his focus from politics to physics, displaying such talent that he became one of Niels Bohr's assistants. As a result, Heisenberg was thoroughly familiar with Bohr's "solar system" model of the atom, in which electrons were viewed as orbiting the nucleus much as planets orbit the sun. At that time, Bohr's model was running into certain theoretical difficulties, and several physicists were trying to resolve them. One was Erwin Schrödinger, whom we have already met. Schrödinger's solution entailed treating the subatomic world as consisting of waves, rather than particles. Heisenberg adopted a different approach. He devised a mathematical system consisting of quantities called matrices (a matrix is a little like a spreadsheet—a rectangular array of numbers arranged in rows and columns) that could be manipulated in such a fashion as to generate known experimental results. Both Schrödinger's and Heisenberg's approaches worked, in the sense that they accounted for more phenomena than Bohr's atomic model. In fact, the two theories were later shown to be equivalent, generating the same results using different ideas.

In 1927, Heisenberg was to make the discovery that would not only win a Nobel Prize, but would forever change the philosophical landscape. Recall that in the late eighteenth century, the French mathematician Pierre Laplace enunciated the quintessence of scientific determinism by stating that if one knew the position and momentum of every object in the universe, one could calculate exactly where every object would be at all future times. Heisenberg's uncertainty principle[6] states that it is impossible to know both exactly where *anything* is and where it is going at any given moment.

These difficulties do not really manifest themselves in the macroscopic world—if someone throws a snowball at you, you can usually extrapolate the future position of the snowball and possibly maneuver to get out of the way. On the other hand, if both you and the snowball are the size of electrons, you're going to have a problem figuring out which way to move, because you will not know where the snowball will go.

We can get a sense of the underlying idea behind Heisenberg's uncertainty principle by looking at an everyday occurrence—the purchase of gasoline at a service station. The cost of the transaction is a number of dollars and cents—the penny is the quantum of our monetary system, the smallest irreducible unit of currency. The cost of the transaction is computed to the nearest penny, and this makes it impossible for us to determine precisely how much gasoline was actually purchased even if we know the exact price per gallon.

If gasoline costs $2.00 per gallon (as it did in the good old days), rounding the cost of the purchase to a penny can result in a difference of $1/200$ of a gallon of gasoline (yes, if you adopt a reasonable rule for rounding, you can cut this to half of $1/200$ of a gallon of gasoline, but the meters in a service station probably round a purchase of $12.5300001 to $12.54). If you start driving from a known position on a straight road and your car gets 30 miles to a gallon, $1/200$ of a gallon of gasoline will take you 0.15 of a mile—792 feet. So the fact that cost is computed in pennies results in a positional uncertainty of 792 feet. I can remember the first time I had the use of a car of my own in the summer of 1961; I used to leave two quarters in the glove compartment for gas in case of emergencies. Gas was about 25 cents a gallon then—at 30 miles per gallon, the cost computed in pennies will result in a positional uncertainty of 1.2 miles. The lower the cost of gasoline, the greater the positional uncertainty. In fact, if gasoline were free, you wouldn't have to pay anything—and you'd have no idea where the car was.

The uncertainty principle operates along similar lines. It states that the product of the uncertainties of two related variables, called conjugate variables, must be greater than some predetermined set amount. Possibly the most familiar conjugate variables are the duration of a musical note and its frequency—the longer the note is held the more accurately we can determine its frequency. A note played for an infinitesimally short period of time simply sounds like a click; its frequency is impossible to determine.

However, the devil in the details of the uncertainty principle comes from the fact that position and momentum (momentum is the product of mass and velocity) are conjugate variables. The more accurately we can determine the position of a particle, the less information we have about its momentum—and if we can determine its momentum to a high degree of accuracy, we have only a limited idea of where it is. Since momentum is the product of mass and velocity, a microscopic quantity of momentum that would amount to almost no velocity at all if allocated to an automobile will result in a lot of velocity if allocated to an electron.

Heisenberg's uncertainty principle is sometimes erroneously interpreted as an *inability* on the part of fallible humans to measure phenomena sufficiently accurately. Rather, it is a statement about the limitations of knowledge, and is a direct consequence of the quantum-mechanical view of the world. As a fundamental part of quantum mechanics, the uncertainty principle has real-world ramifications for the construction of such everyday items as lasers and computers. It has also banished the simple cause-and-effect view of the universe that had been unquestioned since the Greek philosophers first enunciated it. Heisenberg stated one of the consequences of the uncertainty principle as follows:

> It is not surprising that our language should be incapable of describing the processes occurring within the atoms, for, as has been remarked, it was invented to describe the experiences of daily life, and these consist only of processes involving exceedingly large numbers of atoms. Furthermore, it is very difficult to modify our language so that it will be able to describe these atomic processes, for words can only describe things of which we can form mental pictures, and this ability, too, is a result of daily experience. . . . In the experiments about atomic events we have to do with things and facts, with phenomena that are just as real as any phenomena in daily life. But the atoms or the elementary particles themselves are not as real; they form a world of potentialities or possibilities rather than one of things or facts . . . Atoms are not things.[7]

If atoms are not things, what are they? More than seventy-five years after Heisenberg's revelation, physicists—and philosophers—are still struggling with this question. The answer we found previously, that they are probability waves until they are observed, and things thereafter, is not entirely satisfying, but at the moment it is the best we can do.

A Survey in Lower Wobegon

Entanglement, the last of the three conundrums of quantum mechanics we shall investigate, can be translated into a familiar setting. Lower Wobegon, a town located just below Lake Wobegon,[8] differs from Lake Wobegon in that not only are all the children average, the town as a whole is, too—so average that whenever they are polled on a random subject, such as "Do you like asparagus?" 50 percent of the respondents answer yes and 50 percent answer no.

One day, a polling firm decided to sample the opinions of couples in Lower Wobegon. Each pollster was given three questions. Question 1 was "Do you like asparagus?" Question 2 was "Do you think Michael Jordan was the greatest basketball player of all time?" Question 3 was "Do you believe the country is headed in the right direction?"

Two pollsters went into each home. One pollster would ask just one of the three questions of the husband and the other would ask just one of the three questions of the wife—each pollster randomly selecting the questions. Sometimes the questions asked of the husband and the wife were the same, sometimes they were different. When the results were tabulated, 50 percent of the questions were answered affirmatively and 50 percent negatively, but there was something remarkable—when the husband and wife were asked the *same* question, they always answered identically!

Scratching their heads, the pollsters tried to come up with an explanation for this bizarre occurrence. Finally, it occurred to someone that perhaps the husbands and wives all had rehearsed their answers in advance. Even though the questions were not known, they may have formulated a rule such as the following: if the question contains the word *was*, answer yes; otherwise, answer no.

Is there any way to test this hypothesis? Remarkably enough, it can be done. If each husband and wife has formulated such a question-answering rule, there are only four different possibilities—depending upon the three questions, the rule may result in three yeses, or three nos, or two yeses and one no, or two nos and one yes.

Let us look at the responses of the husband and wife to the questions they were asked—even if they were asked different questions (of course, we already know that if they were asked the same question, they responded identically). There are nine different ways the pollsters can ask the three questions. If the rule the husband and wife use results in three yeses or three nos, the husband and wife will always answer their questions identically. If the rule the husband and wife use results in two yeses and one no, let's suppose that the answer to Questions 1 and 2 is yes and to Question 3 is no.

The following table lists all the possibilities.

Husband's Question #	Husband's Response	Wife's Question #	Wife's Response
1	Yes	1	Yes
1	Yes	2	Yes
1	Yes	3	No
2	Yes	1	Yes
2	Yes	2	Yes
2	Yes	3	No
3	No	1	Yes
3	No	2	Yes
3	No	3	No

Notice that in five out of the nine cases (rows 1, 2, 4, 5, and 9 of the table) the answers of the husband and wife agree. When the rule for answering questions produces two yeses and one no, or two nos and one yes, the answers will agree in five out of nine cases. When the rule for answering questions produces three yeses, or three nos, the answers will always agree. *So if husbands and wives have evolved a question-answering rule, it will show up in the data, because when the pollsters go into homes and ask questions of each spouse at random, the husband and wife will produce the same answer at least five times out of nine.*

Convinced that they had the answer to the mystery, the pollsters examined their data. Surprisingly, the answers of husband and wife corresponded approximately half the time. The pollsters concluded that husbands and wives had not evolved a question-answering rule, but this still left a mystery: Why, when they were asked the same question, did the husband and wife always produce the same answer?

Simple, concluded one pollster: when the first spouse was asked a question, he or she communicated which question he or she had been asked and his or her answer; thus, when the second spouse was asked the same question, he or she could answer identically. The solution was simple—prevent the spouses from communicating with each other. These precautions were taken—each spouse was (discreetly) searched for communication devices and questioned in separate rooms. *Still, when each was asked the same question, the answer given by each spouse was the same!*

What could explain this? There are two possibilities that seem to require belief in phenomena not currently addressed by science. The first possibility is that husband and wife possess a sort of intuition—not an explicit means of communicating, but a knowledge of how the other would answer the question. After all, many husbands and wives have the ability to complete each other's sentences.[9] The second possibility is that the marriage is really more than just a uniting, but rather a welding; husband and wife are, in this situation, one. We see the husband and the wife as separate individuals, but with regard to questions asked by pollsters, they are a single entity—to ask a question of one is to ask a question of the other. This differs from the idea of "intuition" in that in the case of intuition, the husband and wife are individual entities who answer the questions identically because they know how the other would answer them. A subtle difference, but a difference nonetheless.

Entanglement and the Einstein, Podolsky, and Rosen Experiment

Many quantum mechanical properties are similar to the wave-particle dilemma faced by photons—until an observation or a measurement is made, the property exists in a superposition of several different possibilities. One such property is the spin around an axis. A photon may spin to the left or to the right around an axis once that axis is selected and the photon observed, but it will spin to the left 50 percent of the time and to the right 50 percent of the time, and do so randomly. This is clearly similar to the responses to poll questions of the inhabitants of Lower Wobegon.

When a calcium atom absorbs energy and later returns to its initial state, it emits two photons whose properties parallel the responses to poll questions of husbands and wives in Lower Wobegon. The photons are said to be entangled—the result of the measurement of the spin of one of the photons automatically determines the result of the measurement of the spin of the other, even though initially neither photon possesses spin, but only a probability wave that allows for left and right spin equally. At least, that is a viewpoint that is largely accepted by physicists.

Albert Einstein was extremely uncomfortable with this point of view, and he and physicists Boris Podolsky and Nathan Rosen devised, in 1935, a thought experiment, known as the EPR experiment,[10] that challenges the idea. Einstein, Podolsky, and Rosen objected to the concept that before the measurements, neither spin is known. Suppose two groups of experimenters, light-years apart, set out to measure the spins of these photons. If the spin of photon A is measured, and seconds later the spin of photon B is measured, quantum mechanics predicts that photon B would "know" the result of the measurement on photon A, even though there would not be enough time for a signal from photon A to reach photon B and tell photon B what its spin should be!

According to Einstein, this left two choices. One could accept the so-called Copenhagen interpretation of quantum mechanics, due primarily to Niels Bohr, that photon B knows what happened to photon A even without a signal passing between them. This possibility, corresponding to "intuition" in Lower Wobegon, is doubtless the reason that quantum mechanics seems to open the door to mysticism in the real world. After all, what could be more mystical than knowledge of what happens to another body without a measurable transmission of information? Alternatively, one could believe that there is a deeper reality, manifested in some physical property as yet unfound and unmeasured, which would account for this phenomenon—this corresponds to "rehearsed answers" in Lower Wobegon. Einstein died holding firmly to this latter view, which is known in the physics community as "hidden variables."

Bell's Theorem

More than a hundred papers were written between 1935 and 1964 discussing the pros and cons of the hidden variables explanation, but these were just discussions and arguments—until the Irish physicist John Bell came up with a highly ingenious experiment that would subject the hidden variables theory to an actual test. Bell suggested that the experiment should consist of an apparatus that could measure the spin of each photon around one of three axes. The axis for each photon would be randomly selected, and the spins of the two photons recorded. These measurements would be recorded as pairs: the pair (2,L) indicates that axis 2 was selected for measurement and the photon was spinning left around this axis.

Suppose that the two entangled photons are each imprinted with the following program: if axis 1 or axis 2 is selected, spin to the left; if axis 3 is selected, spin to the right. Assuming the axis for each photon is randomly selected, there are nine possible choices of axes, just as there were nine possible choices of questions for the two pollsters in Lower

Wobegon. Axial spin is another example of conjugate variables—it is impossible to simultaneously determine the spin of a photon around more than one axis. This parallels the situation in Lower Wobegon as well—each pollster only asked a single question.

Bell devised this as a thought experiment to test whether the hypothesis that the photons had a hidden program imprinted on both, as suggested by Einstein, Podolsky, and Rosen, was valid. If such were the case, the two photons would have the same spin more than $5/9$ of the time. Within a few years, thousands of trials of Bell's experiment were performed—*and the detectors did not record the same direction of spin for the photons more than half the time.* This constituted undeniable proof that there was no hidden program with which the photons were imprinted.

Was there another possible solution? With the hidden variables explanation ruled out, the next most likely possibility was that somehow the photons could signal each other. The instant the first photon's spin was recorded, it sent a message to the other photon along the lines of "Someone just measured my spin around axis 1 and I spun to the left."

The theory of relativity places no restrictions on the existence of signaling mechanisms, but it does require that no signals can be sent faster than the speed of light. By the early 1980s, improvements in technology enabled a more sophisticated version of the above experiment to be performed. In this experiment, the detector equipment for the two photons were placed a significant distance apart, and a randomizing device installed that would select the axis for the second photon *after* the spin of the first photon had already been measured. What made this experiment so interesting was that a new wrinkle had been added: the technology was now so good that the axis for the second photon could be selected in a shorter period of time than it would take a light beam to go from the first detector to the second. The first photon to be measured could therefore send a signal to the other photon, but it could not be received in time for the second photon to act upon it—the spin of the second photon was measured before a signal traveling at the speed of light from the first photon could reach it. The results of this experiment were obtained in a laboratory by Alain Aspect[11] in 1982 with a detector separation distance measured in the meters. The separation distance was increased to eleven kilometers in the late 1990s, but the results were still the same. If the same axis were chosen, the photons always spun in the same direction—but they spun in the same direction no more than half the time.

This is one of the great mysteries of quantum mechanics, unresolved more than a century after Max Planck first hatched the idea of the quantum. Although special relativity forbids either matter, energy, or information from traveling faster than light, what is happening here is that the probability wave has collapsed instantly throughout the entire universe.

There is a dramatic moment in the first *Star Wars* movie (Episode Four) when Obi-Wan Kenobi senses a great disturbance in the Force. You don't have to be Obi-Wan Kenobi to sense a disturbance in a probability wave; the universe does it for you by collapsing it instantaneously and everywhere when an observation is made.

But how is this done? At present there are suggestions and ideas— including the idea that this is something we may never know. Even if we never know it, the pursuit of this knowledge will undoubtedly result in developments, both technological and philosophical, that will greatly change our world. Sir Arthur Eddington, who led the 1919 expedition that confirmed Einstein's theory of relativity, may have put it best when he said, "The Universe is not only stranger than we imagine, it is stranger than we can imagine"[12]—because who could ever have imagined wave-particle duality, the uncertainty principle, and entanglement?

Round 1

Samuel Johnson had his Boswell, but John Wheeler certainly deserves one—possibly no physicist or mathematician is able to encapsulate the dilemmas faced by science so succinctly. A key component of the incompatibility between the physics of the large (relativity) and the physics of the small (quantum mechanics) is the mathematical model used to describe it. At the level of the really, really small, the hands-down winner—for the moment—is the discrete view, because Max Planck's hypothesis resulted in discrete descriptions that were unbelievably successful in predicting values of all the relevant physical quantities. This triumph would have come as a vindication to a small band of quasi-religious mystics who lived two and a half millennia ago, and whom we will encounter in the next chapter.

NOTES

1. See http://en.wikipedia.org/wiki/Niels_Bohr. Go to this site for the bio, but stay for the quotes. Neils Bohr is part Yogi Berra, part Yoda. Here's a teaser quote, which should be studied assiduously by every public figure: "Never talk faster than you think."
2. See http://en.wikipedia.org/wiki/Rayleigh-Jeans_Law. This brief site is tremendously intriguing, as it has the equations for both the Rayleigh-Jeans law and Planck's revision, as well as an attractive graphic that illustrates the ultraviolet catastrophe.

3. J. Bronowski, *The Ascent of Man* (Boston: Little, Brown, 1973), p. 336.

4. See http://en.wikipedia.org/wiki/Schr%C3%B6dinger%27s_cat. Physics is replete with tremendously provocative thought experiments. This site has a pretty thorough discussion.

5. R. Hillmer and P. Kwiat, "A Do-It-Yourself Quantum Eraser," *Scientific American,* May 2007. However, if in building this you accidentally erase the entire universe, neither I nor the publisher can be sued.

6. See http://en.wikipedia.org/wiki/Uncertainty_Principle. This site has a good derivation if you know linear algebra and the Cauchy-Schwarz inequality; this is usually upper-division mathematics and physics.

7. W. Heisenberg, *The physical principles of the quantum theory* (Chicago: University of Chicago Press, 1930).

8. A fictitious town, invented by Garrison Keillor and described on his National Public Radio show, *A Prairie Home Companion,* as a town where "the women are strong, the men are good looking, and all the children are above average." The Lake Wobegon effect, in which everyone claims to be above average, has been observed in automobile drivers and college students (in estimating their mathematical abilities).

9. Remarkably, though, surveys have shown that a far greater proportion of wives complete their husbands' sentences than vice versa.

10. See http://en.wikipedia.org/wiki/EPR_experiment. This is an excellent site, and also has material on Bell's inequality, thus saving me a search.

11. See http://www.drchinese.com/David/EPR_Bell_Aspect.htm. If, as the TV show *Mr. Ed* put it, you want to go "right to the source and ask the horse," this site enables you to download in PDF format the Big Three papers in this area (the EPR experiment, Bell's theorem, and the Aspect experiment). All three basically require high-level degrees, but if you want to see the original versions, here they are. It also has photos of the three main protagonists—you might mistake Geraldo Rivera for Alain Aspect.

12. See http://www.quotationspage.com/quote/27537.html.

Section II

The Incomplete Toolbox

4

Impossible Constructions

The Brotherhood

It was a powerful secret society of men bound together by common beliefs in religion and mysticism. Then, one day, their entire belief structure would be shattered by a discovery so profound it would transform the thinking of the civilized world.

It sounds like a description of Opus Dei, the powerful clandestine Catholic secret society that played a pivotal role in the blockbuster novel *The Da Vinci Code*. Alternatively, it might have described the core of the church during the seventeenth century, when it was confronted by Galileo's shattering discovery that the moons of Jupiter orbited a celestial body other than Earth. This secret society, however, existed some two millennia prior to Galileo. Founded by the philosopher-mathematician Pythagoras, the motto of the society—"All is number"—reflected the view that the universe was constructed of either whole numbers or their ratios. The discovery that was to rock their world was that the square root of 2, the ratio of the length of the diagonal of a square to its side, was incommensurable—that is, it could not be expressed as the ratio of two whole numbers.

The Greeks actually constructed both numerical and geometrical proofs of this fact—the numerical proof was based on the concept of odd and even numbers. If the square root of 2 could be expressed as the ratio p/q of whole numbers, those numbers could be chosen to have no common factor (we learned in elementary school to cancel common factors to reduce fractions). If $p/q = \sqrt{2}$, then $p^2/q^2 = 2$, and so $p^2 = 2q^2$. Since p^2 is a multiple of 2, p must be an even number, as odd numbers have odd squares. Since p and q have no common factor, q must be odd. Letting $p = 2n$, we see that $(2n)^2 = 2q^2$, and so $q^2 = 2n^2$; the same reasoning as we used to show that p is even shows that q is even—and we have thus concluded that q is both even and odd.

The discovery of the incommensurability of the square root of 2 affected the development of Greek mathematics as profoundly as the discovery of the moons of Jupiter affected the development of astronomy. The Greeks turned from the philosophy of *arithmos* (the belief in number that is obviously the root of our word *arithmetic*) to the logical deductions of geometry, whose validity were assured.

The geometry of the Greeks—later to be formalized by Euclid—was initially based on the line and the circle. The tools for exploring geometry were the straightedge, for drawing lines and line segments, and the compass, for creating circles. There does not seem to be a record as to why the Greeks required that the straightedge be unmarked, so that no distances were inscribed upon it. Possibly the earliest Greek geometers had access only to the simplest tools, and the use of compass and unmarked straightedge simply became the traditional way to do geometry. However, it wasn't until the Greeks began the exploration of figures other than those constructed with lines and circles that the utility of the marked straightedge began to reveal itself; it brought a certain ungainly aspect into geometrical constructions, but greatly increased their scope. That exploration did not begin until four hundred years before the birth of Christ, and after another profound event was to shake the foundations of ancient Greece.

The First Pandemic

In 430 BC, the Athenians were engaged in the Peloponnesian War when a plague overtook the city. The historian Thucydides was taken ill but survived, and described the horrifying course of the disease.[1] The eyes, throat, and tongue became red and bloody, followed by sneezing, coughing, diarrhea, and vomiting. The skin was covered in ulcerated sores and pustules, accompanied by a burning, unquenchable thirst. The disease started in Ethiopia, and spread to Egypt, Libya, and then to Greece. The plague lasted for almost four years, and killed a third of the Athenian population. Only recently have we discovered through DNA analysis that the disease was actually typhoid fever.[2]

One can only imagine the desperation of the people, who were almost certainly willing to attempt anything that had even the remotest chance of alleviating the devastation. The oracle at Delos was consulted, and the recommended remedy was to double the size of the existing altar, which was in the shape of a cube.

It was easy to double the edge of a cube, but this would have created an altar with a volume eight times the size of the initial one. The Greeks were highly skilled in geometry, and realized that in order to construct a cube whose volume was double the size of the initial one, the edge of the doubled cube would have to exceed the length of the original cube by a factor of the cube root of 2. None of the sages could use these instruments to construct an edge of the desired length using only the compass and unmarked straightedge. Eratosthenes relates that when the craftsmen, who were to construct the altar, went to Plato to ask how to resolve the problem, Plato replied that the oracle didn't really want an altar of that size, but by so stating the oracle intended to shame the Greeks for their neglect of mathematics and geometry.[3] In the midst of a plague, receiving a lecture on the mathematical deficiencies of Greek education was probably not what the craftsmen or the Athenian populace wanted to hear.

It took four years for the plague to burn out, but the problem of constructing a line segment of the desired length endured—either because the Greeks relished the intellectual challenge of the problem, or as a possible defense against a recurrence of the plague. At any rate, the problem of constructing a line segment of the desired length was solved by several different mathematicians using a variety of approaches.

Probably the most elegant of the solutions was that proposed by Archytas, who constructed a solution based on the intersection of three surfaces: a cylinder, a cone, and a torus (a torus looks like the inner tube of a tire). This solution demonstrated a considerable amount of sophistication—solid geometry is considerably more complex than plane geometry (I took a course in solid geometry in high school and received a B minus; to this day, it remains one of the toughest math courses I've ever taken). Two simpler solutions were found by Menaechmus using plane curves: the intersection of two parabolas, and the intersection of a hyperbola and a parabola.[4]

Archytas's and Menaechmus's solutions are representative of a theme we shall see throughout this book—the quest for solutions to a problem, even an impossible one, often leads to fruitful areas where no man, or mathematician, has gone before. Menaechmus is credited with the discovery of the hyperbola and parabola,[5] which are two of the four conic sections, the others being the circle and the ellipse. Each of the curves is the intersection of a plane and a cone, and each of the curves not only recurs constantly in nature, but has also been incorporated in many of the devices that characterize our technological age: the parabola in the parabolic reflectors of satellite dishes; the ellipse in lithotripsy machines, which break up kidney stones using sound waves rather than invasive surgery; and the hyperbola in navigational systems such as loran.

The Greeks did more than simply discover solutions to mathematical problems; they also used them. Plato invented a device known as Plato's machine, which utilized geometry, to construct a line segment whose length was the cube root of the length of a given line segment. Plato, however, was not the only savant to tackle the physical construction of doubling the cube. Another person to undertake this task was Eratosthenes. His construction, involving simple rotations of lines and attached triangles, was capable of being adapted to construct not only cube roots, but any integer root. Eratosthenes supplied color commentary on his construction, complete with a denigration of rival techniques.

"If, good friend, thou mindest to obtain from any small cube a cube the double of it, and duly to change any solid figure into another, this is in thy power; thou canst find the measure of a fold, a pit, or the broad basin of a hollow well, by this method, that is, if thou thus catch between two rulers two means with their extreme ends converging. Do not thou seek to do the difficult business of Archytas's cylinders, or to cut the cone in the triads of Menaechmus, or to compass such a curved form of lines as is described by the god-fearing Eudoxus."[6]

Eratosthenes's self-serving remarks may have been prompted by his being given the nickname Beta (the second letter of the Greek alphabet) by his contemporaries, who felt that his not-inconsiderable achievements (among which were the first accurate measurement of the circumference of Earth, the compilation of a star catalog, and numerous contributions to mathematics, astronomy, and geography) never merited the supreme accolades given to the best of the best.

"[Eratosthenes] was, indeed, recognised by his contemporaries as a man of great distinction in all branches of knowledge, though in each subject he just fell short of the highest place. On the latter ground he was called Beta, and another nickname applied to him, Pentathlos, has the same implication, representing as it does an all-round athlete who was not the first runner or wrestler but took the second prize in these contests as well as others."[7] It seems that even in ancient Greece, they felt that another name for the runner-up is "loser."

Other problems in geometry, although not appearing to have consequences as significant as plague prevention, perplexed the Greek mathematicians. Two of these problems, squaring the circle and trisecting the angle, were solved by the Greeks by reaching outside the realm of classical straightedge-and-compass constructions. The third problem, the construction of regular polygons (a polygon is regular if all its sides are the same length and if all the angles formed by adjacent sides are equal—the square and the equilateral triangle are regular) with an arbitrary number of sides, eluded them.

The term *squaring the circle*—constructing a square whose area is the same as that of a given circle—is often used as a shorthand for an impossible task. As with doubling the cube, the task was not impossible. Archimedes described a neat construction that began by "unrolling" the given circle to produce a line segment whose length was the circumference of the given circle.[8] However, unrolling is not a straightedge-and-compass construction. Similarly, the task of trisecting the angle—constructing an angle whose degree measure is one-third the degree measure of a given angle—can easily be accomplished by making a mark on the straightedge that is being used (a method that is also ascribed to Archimedes), which also lies outside the framework of classical straightedge-and-compass constructions allowed in Euclidean geometry. These constructions showed that the Greeks, though recognizing the formal restrictions of Euclidean geometry, were willing to search for solutions to problems even if those solutions could only be found outside the system in which the problems were posed.

We do not know if the Greeks ever conjectured that these tasks could not be accomplished within the framework of straightedge-and-compass constructions. It is certainly easy to believe that a mathematician such as Archimedes, having expended a good deal of effort on one of these problems, might well have reached such a conclusion. What we do know is that even today, when the impossibility of such tasks has been proven to the satisfaction of at least five generations of mathematicians, countless man-hours are spent formulating "proofs" and sending the results to mathematical journals. Some of the people who spend time on these problems are unaware that mathematicians have proved that trisecting the angle or squaring the circle is impossible.[9] Others are aware, but either believe that mathematical impossibility is not an absolute, or the proof of that impossibility is flawed.

There are simple straightedge-and-compass constructions to construct regular polygons with three, four, and six sides, and there is a slightly more complex construction for a regular polygon with five sides. All these were known to the ancient Greeks, but a classical construction of other regular polygons proved elusive.[10] In the late 1920s, a manuscript attributed to Archimedes (who else?) was discovered that outlined a method of constructing a regular heptagon by sliding a marked straightedge, but almost two millennia would elapse from the era of Archimedes before the four problems under discussion were finally resolved to the satisfaction of the mathematical community.

The Mozart of Mathematics

Any list of the greatest mathematicians must include Carl Friedrich Gauss (1777–1855), the Mozart of mathematics, whose mathematical talents were evident at an extraordinarily young age. At the age of three, he was presumably studying his father's accounts, and correcting arithmetic errors if and when they occurred. Just as Mozart is renowned for having composed music at an exceptionally young age, Gauss is also known for demonstrating genius at an early age. During an arithmetic lesson in elementary school, the class was asked to add the numbers from 1 to 100. Gauss almost immediately wrote "5050" on his slate, and exclaimed, "There it is!" The teacher was stunned that a child could find the correct answer so quickly; the technique Gauss employed is still known as "the Gauss trick" to mathematicians. Gauss realized that if one wrote down the sum

$$S = 1 + 2 + 3 + ... + 98 + 99 + 100$$

and then wrote down the same sum in reversed order

$$S = 100 + 99 + 98 + ... + 3 + 2 + 1$$

if one were to add the left sides one would get $2S$, and if one were to add the right side by thinking of it in terms of 100 pairs of numbers, each of which summed to 101 ($1 + 100$, $2 + 99$,..., $99 + 2$, $100 + 1$), one would obtain $2S = 100 \times 101 = 10,100$, and so $S = 5050$.[11]

Even more incredible is the fact that when Gauss was given a table of logarithms at age fourteen, he studied it for a while, and then wrote on the page that the number of primes less than a given number N would approach N divided by the natural logarithm of N as N approached infinity. This result, one of the centerpieces of analytic number theory, was not proved until the latter portion of the nineteenth century. Gauss did not supply a proof, but even to be able to conjecture this at the age of fourteen is simply extraordinary.[12]

When he was nineteen years old, Gauss supplied a straightedge-and-compass construction for the regular heptadecagon—the polygon with seventeen sides. Moreover, his construction technique showed that polygons with $2^{2^N} + 1$ sides were regular (numbers of this form are known as Fermat primes,[13] as they were first studied by the French mathematician Pierre de Fermat, of Fermat's last theorem fame). More than two thousand years had elapsed since anyone had shown constructions of regular polygons other than the constructions known to the ancient Greeks.

A list of Gauss's accomplishments would take substantial time and space—suffice it to say that his career fulfilled its early promise. He is recognized today as one of the two or three greatest mathematicians of all time—and this does not even include his noteworthy accomplishments in the fields of physics and astronomy.

Pierre Wantzel: The Unknown Prodigy

I'm not a mathematical historian, and at the time of the writing of this book, the name of Pierre Wantzel was unfamiliar to me, and I suspect it would be equally unfamiliar to many of today's mathematicians. Wantzel was born in 1814, the son of a professor of applied mathematics. Like Gauss, his talent for mathematics manifested itself at an early age—where Gauss was correcting errors in his father's accounts, Wantzel was handling difficult surveying problems when he was only nine years old. After a brilliant academic career in both high school and college, Wantzel entered engineering school. However, feeling that he would experience greater success teaching mathematics than doing engineering, he became a lecturer in analysis at the École Polytechnique—at the same time that he was a professor of applied mechanics at another college, while also teaching courses in physics and mathematics at other Parisian universities.

Gauss had stated that the problems of doubling the cube and trisecting the angle could not be solved by straightedge-and-compass construction, but he had not supplied proofs of these assertions. This was standard operating procedure for Gauss in many problems, but it sometimes left his colleagues in a dilemma as to whether they should work on a particular problem, only to find that Gauss had previously solved it. Wantzel, however, was the first to publish proofs of Gauss's assertions—finally laying to rest these two problems. Wantzel had also simplified the proof of the Abel-Ruffini theorem on the roots of polynomials, and used this to show that an angle was constructable if and only if its sine and cosine were constructable numbers. Simple trigonometry showed that the sine and cosine of a 20-degree angle were not constructable numbers.[14] Additionally, Wantzel polished off the problem of which regular polygons were constructable by showing that the only such regular polygons were those with n sides, where n is a product of a power of 2 and any number of Fermat primes.

Jean Claude Saint-Venant, who was one of the leading French mathematicians of the period and a colleague of Wantzel's, described his habits as follows: "He usually worked during the evening, not going to bed until late in the night, then reading, and got but a few hours of agitated sleep, alternatively abusing coffee and opium, taking his meals, until his marriage, at odd and irregular hours." Saint-Venant further commented upon Wantzel's failure to achieve more than he had (even though his achievements would do credit to 99 percent of the mathematicians who have ever lived) by further stating, "I believe that this is mostly due to the irregular manner in which he worked, to the excessive number of occupations in which he was engaged, to the continual movement and feverishness of his thoughts, and even to the abuse of his own facilities."[15]

The Impossibility of Squaring the Circle

By the middle of the nineteenth century, the lengths of line segments that could be constructed had been shown to be the result of applying addition, subtraction, multiplication, division, and the taking of square roots to integers (since cube roots cannot be obtained by this process, the cube could not be doubled nor the angle trisected). In order to square a circle of unit radius, since the area of the circle is π, one must be able to construct a line segment whose length is the square root of π, which can only be done if one can construct a line segment whose length is π.

By this time, mathematicians had shown that the real line consisted of two types of numbers: the rational numbers such as $22/7$, which could be viewed as either the quotient of two integers or the ratio of one integer to another, and the irrational numbers, those which could not be expressed as ratios. As we have seen, the Pythagoreans knew that the square root of 2 is irrational; this knowledge was so well known to educated Greeks that a proof of it appears in one of the Socratic dialogues.[16] The irrational numbers had been further subdivided into the algebraic numbers, those numbers that were the roots of polynomials with integer coefficients, and the transcendental numbers. In 1882, the German mathematician Ferdinand von Lindemann wrote a thirteen-page paper showing that π was transcendental, thus showing that the circle could not be squared by straightedge-and-compass construction. To this day Lindemann gets the credit, although much of the earlier work was done by the French mathematician Charles Hermite. Lindemann's proof of the transcendentality of π is similar to Hermite's proof that e, the base of the natural logarithm, is transcendental. In the nineteenth century, fame was the only reward for a mathematician—although today there are monetary prizes offered as incentives for the solutions of major problems. Then, as now, fame generally accrued to the person who placed the final brick in place on the edifice, rather than those who laid the foundation.[17]

Learning from Impossibility

All of the problems investigated in this chapter are great problems. A great problem is generally relatively simple to explain, piques our curiosity, is difficult to resolve, and has a resolution that extends the bounds of what we know beyond the problem itself. It makes us question whether the assumptions we have made are sufficient to solve the problem, and whether the tools we have are adequate for the job.

The quest to double the cube and trisect the angle led to explorations far beyond the simple structures of line and circle that constitutes Euclidean plane geometry. The axioms of plane geometry, as given in the first book of Euclid's *Elements*, are

1. Any two points can be joined by a straight line.
2. Any straight line segment can be extended indefinitely in a straight line.
3. Given any straight line segment, a circle can be drawn having the segment as radius and one endpoint as center.
4. All right angles are congruent.
5. (Parallel postulate) If two lines intersect a third in such a way that the sum of the inner angles on one side is less than two right angles, then the two lines inevitably must intersect each other on that side if extended far enough.[18]

The above postulates discuss only points, lines, angles, and circles. Even though an outline of the *Elements* reveals both plane and solid geometry, the geometrical figures that are discussed are polygons and polyhedra, circles and spheres. The duplication methods proposed by Archytas, Menaechmus, and Eratosthenes certainly transcend Euclidean geometry as outlined in the *Elements*.

Attempts to square the circle led to a deeper analysis of the real line and the concept of number. The resolution of the problem of constructing regular polygons revealed a surprising connection between geometry and an interesting class of prime numbers. Indeed, this is one of the persistently surprising and appealing aspects of mathematics—there are unexpected connections between not only areas of mathematics, but also between mathematics and other areas.

However, mathematics sometimes causes people to leap to unsupported conclusions. In trying to fit the orbits of the planets into a coherent pattern, Johannes Kepler was struck by the coincidence that, at the time, there were six planets and five regular solids. More planets would be discovered, but Greek mathematicians had proved that there were only five regular solids: the four-sided tetrahedron, the six-sided cube, the eight-sided octahedron, the twelve-sided dodecahedron, and the twenty-sided icosahedron. Based on inadequate data, Kepler constructed this model.

"The Earth's orbit is the measure of all things; circumscribe around it a dodecahedron, and the circle containing this will be Mars. Circumscribe around Mars a tetrahedron, and the circle containing this will be Jupiter. Circumscribe around Jupiter a cube, and the circle containing this will be Saturn. Now inscribe within the Earth an icosahedron, and the circle contained within it will be Venus. Inscribe within Venus an octahedron, and the circle contained within it will be Mercury. You now have the reason for the number of planets."[19]

An exquisitely beautiful scheme—but dead wrong. The lure of pattern is so strong that just as we think we see a face on Mars when it is merely a land formation seen in light that accentuates features that appear to be human, we sometimes see mathematical patterns based on inadequate data or information. It is to Kepler's eternal credit that he did something that must have been extremely difficult: when Tycho Brahe supplied him with better data to which he could not get the model to fit, he abandoned the model. In so doing, he formulated Kepler's laws of planetary motion, which led to Newton's discovery of the theory of universal gravitation.

Pythagoreans Redux

The fundamental tenet of the Pythagoreans was that the universe was constructed of whole numbers and the ratios of whole numbers. The discovery that the square root of 2 was incommensurable destroyed this worldview—at the time of the Pythagoreans. However, in an intriguing twist, the Pythagoreans may have been right after all! Quantum mechanics, so far the most accurate depiction of the universe that we have, is essentially a modern version of the view espoused by the Pythagoreans. As we have seen, the world according to quantum mechanics consists of a collection of whole numbers of basic units—mass, energy, length, and time are all measured in terms of quanta. The mathematics of the real number system, the residence of the square root of 2, is an ideal construct that has great utility and considerable intellectual interest. In the real world, though, if a square can actually be constructed out of material objects, its diagonal (also constructed out of material objects) is either a little too short to reach from corner to corner or extends a little beyond.

It makes one wonder if other ideas, known to previous civilizations but long since discarded, are waiting quietly on the sidelines to make a comeback in modern guise.

NOTES

1. http://www.perseus.tufts.edu/GreekScience/Thuc.+2.47-55.html.
2. *International Journal of Infectious Diseases,* Papagrigorakis, Volume 11, 2006.
3. T. L. Heath, *A History of Greek Mathematics I* (New York: Oxford, 1931).
4. http://www-groups.dcs.st-nd.ac.uk/~history/HistTopics/Doubling_the_cube .html#s40. This gem of a Web site contains not only Archytas's and Menaechmus's solutions to duplicating the cube, but also Eratosthenes' method of finding roots. A certain comfort level with analytic geometry is required to stay up with Archytas, but Menaechmus's solutions are fairly straightforward and a high-school graduate shouldn't have much difficulty with them. Even if the reader doesn't intend to "do the math" necessary to follow the constructions, the site is worth looking at simply to gain a greater appreciation of the sophistication of the ancient Greeks. The fact that they could do all these things using only geometry (no analytic geometry, which greatly simplifies all things geometrical) and not having access to pencil and paper, still causes me to shake my head in disbelief—and we haven't even gotten to Archimedes.
5. T. L. Heath, *A History of Greek Mathematics I* (New York: Oxford, 1931).
6. Ibid.
7. Ibid.
8. A. K. Dewdney, *Beyond Reason* (New York: John Wiley & Sons, 2004), p. 135. This construction is by no means Archimedes' finest hour—but an off day for Archimedes would make the career of many lesser mathematicians. He simply uses the length of the unrolled circumference as the base of a right triangle, and the radius of the circle as the height of that triangle. This results in a triangle whose area is $1/2 \times (2\pi r) \times r = \pi r^2$, and a standard construction will produce a square with the same area as the triangle.
9. http://www.jimloy.com/geometry/trisect.htm. This site probably sets the record for most erroneous trisections of an angle—some extremely ingenious and only subtly in error. I wish it had been in existence when I was a junior faculty member at UCLA back in the late 1960s. Then as now, UCLA was the home of the *Pacific Journal of Mathematics.* Back then it would receive numerous submissions for trisection of the angle—and the editors, generally a polite group, would respond not with a curt, "It's impossible, don't bother sending anything else,"

but with a detailed analysis of the error in the "proof." And guess who got to perform that analysis? Junior faculty members—like me. I learned a lot of geometry tracing down those errors, but I could have saved a lot of time if this site had been accessible.

10. http://mathworld.wolfram.com/GeometricConstruction.html. This site has a reasonable version of how to construct equilateral triangles, squares, regular pentagons, and Gauss's heptadecagon. Mathworld has a lot of good stuff—some of it is highly technical but it's always a reasonable place to start. I think it's mostly written either to sell or support Mathematica, a Wolfram product, which many mathematicians swear by. As a result, it sometimes reads a little like something in Math Reviews.

11. http://en.wikipedia.org/wiki/Carl_Friedrich_Gauss. I admit that Wikipedia has encountered some problems—since anyone is free to edit it, sometimes the individual editing Wikipedia has an agenda and uses Wikipedia to promulgate that agenda. However, that rarely happens with mathematics—it's hard to imagine anyone having an agenda about Carl Friedrich Gauss, the early years. Additionally, there are often a lot of secondary references in Wikipedia that can be used either to pursue a subject in more depth or to authenticate the material.

12. http://www.math.okstate.edu/~wrightd/4713/nt_essay/node17.html. This site not only has Gauss' original conjecture, but several related ones. It helps to know a little calculus, but most of what is said requires only the knowledge of what natural logarithms are.

13. http://en.wikipedia.org/wiki/Fermat_number#Applications_of_Fermat_numbers. There is an attractive theorem that states that if $2^n + 1$ is a prime, then n must be a power of two. Fermat primes continue to be studied; a recent intriguing result is that no Fermat number can be the sum of its divisors. Numbers such as $6 = 1 + 2 + 3$ and $28 = 1 + 2 + 4 + 7 + 14$, which are the sum of their divisors, are called perfect numbers. Fermat numbers are also useful for generating sequences of random integers for use in computer simulations.

14. http://planetmath.org/encyclopedia/TrisectingTheAngle.html. This site contains a plethora of material on this and related problems.

15. http://www-groups.dcs.st-and.ac.uk/~history/Biographies/Wantzel.html. This site has a number of good biographies, including the best easy-to-find one of Wantzel.

16. The actual dialogue is Plato's *Meno*, in which Socrates works with an unschooled servant boy to discover that the square root of 2 is irrational. A good account of the argument can be found in http://www.mathpages.com/home/kmath180.html. Although philosophy was not my best subject in high school (or anywhere else), I remember my instructor telling us that there was a lot of unconscious humor in *Meno*. Socratic dialogues were sometimes the equivalent of a paid appearance by Paris Hilton—Socrates was paid to conduct a dialogue as entertainment for the guests at a banquet. The subject of this dialogue was "virtue"—a sly dig at Meno, who my philosophy instructor told us was something of the Godfather of his day.

17. http://www-groups.mcs.st-and.ac.uk/history/Biographies/Lindemann.html. As mentioned previously, this site has numerous good biographies and secondary references. It also has terrific internal hyperlinking, so you can jump around and get a lot of related information.

18. One of the great benefits of living in the Information Age is the extraordinary amount of classic material that is available online. Here is a wonderful version of Euclid's classic work, complete with useful Java applets. http://aleph0.clarku .edu/~djoyce/java/elements/toc.html.

19. http://www.astro.queensu.ca/~hanes/p014/Notes/Topic_019.html. The quote appears in the section entitled "Kepler the Mystic."

5 The Hope Diamond of Mathematics

The Curse

The Hope Diamond is probably the most famous diamond in the world. Its fame comes not so much from its size—at 45.52 carats, it's certainly impressive, but the Kohinoor Diamond tips the scales at 186 carats—nor from its brilliant blue color, which is due to traces of boron impurities. Its fame is based on the belief that all who possess it will suffer a curse inflicted by the Hindu goddess Sita, who will exact revenge because the diamond was originally the eye of an idol dedicated to her, and was stolen from it.

Legend[1] has it that a jeweler named Tavernier stole the diamond originally, and was torn to death by wild dogs on a trip to Russia. The Hope Diamond was owned at one time by Louis XVI and Marie Antoinette, who were beheaded during the French Revolution. The diamond's name comes from one of its owners, Henry Thomas Hope, whose grandson Henry Francis Hope gambled and spent his way to bankruptcy. The Hope (it is customary to refer to famous diamonds by name only) was eventually purchased by Evalyn Walsh McLean—whose riches could purchase diamonds, but not forestall tragedy. Her firstborn son died at age nine in a car crash, her daughter committed suicide when she was twenty-five, and her husband was declared insane and lived out his life in a mental institution.

The Hope left a swath of misfortune in its wake—but this pales in comparison to the sufferings of the major players in the search for the solution to polynomial equations of ever-higher degree.

A Mathematician's Job Interview

Mathematics students are told of the mathematician who applied for a job with a corporation. When asked what he could do, the mathematician replied that he solved problems. The interviewer took him to a room in which a fire was blazing. There was a table on which rested a bucket of water, and the mathematician was instructed to put out the fire. The mathematician grabbed the bucket, dumped water on the fire, and extinguished it. He then turned to the interviewer and asked, "Do I get the job?"

"You'll have to take the advanced test," replied the interviewer. The mathematician was taken to another room in which a fire was blazing. There was a table, under which rested a bucket of water, and the mathematician was instructed to put out the fire. The mathematician grabbed the bucket—and placed it on top of the table. Why on Earth, students want to know, would he do that? Because mathematicians like to reduce a new problem to one they've previously solved.

Progress in mathematics is often cumulative, with previous results being used to derive ever deeper and more complex results. Such is the story of the search for the solutions of polynomial equations, such as $ax^3 + bx^2 + cx + d = 0$, which is the general polynomial equation of degree 3. Polynomials are the only functions that we can calculate,[2] because they involve only addition, subtraction, multiplication, and division. With rare exceptions, when we calculate a value such as a logarithm or the sine of an angle (for example, by using a calculator), the logarithm or sine is approximated by a polynomial, and it is this approximate value that is calculated.

Early Results: Solutions of Linear and Quadratic Equations

The story of the search for the solution to polynomial equations starts tamely enough in ancient Egypt, whose mathematicians were sufficiently adept to solve linear equations. An example of one such is the equation $7x + x = 19$, which nowadays is comfortably solved by sixth-graders, who would add the terms on the left to obtain $8x = 19$, and then divide both sides by 8 to show that $x = {}^{19}/_8$. Algebra was unavailable to Ahmes of Egypt, who contributed a section entitled "Directions for Knowing All Dark Things" (many modern-day students would doubtless agree with this definition of mathematics) to the Rhind papyrus, one of the first mathematical manuscripts. Ahmes solved this problem using a method that can only be described as tortuous.[3]

The Hanging Gardens may have been the only physical contribution of Babylonians to the wonders of the ancient world, but their mathematical accomplishments were quite impressive for the times. They were capable of solving certain quadratic equations (equations of the form $ax^2 + bx + c = 0$) using the method of completing the square,[4] which is used in high-school algebra to generate the full solution to this equation. The resulting formula is known as the quadratic formula. It was described early in the ninth century by the Arab mathematician Al-Khowarizmi, who is also responsible for giving the name *algebra* to algebra.

Del Ferro and the Depressed Cubic

Time passed—approximately seven centuries. There would be no significant advance in equation solving until the middle of the fifteenth century, when a collection of brilliant Italian mathematicians embarked upon a quest to solve the equation $ax^3 + bx^2 + cx + d = 0$. This equation, which is known as the general cubic, was to prove a far tougher nut to crack.

As the degree of the polynomial increases, different types of numbers are needed to solve it. Equations such as $2x - 6 = 0$ can be solved with positive integers, but $2x + 6 = 0$ requires negative numbers, and $2x - 5 = 0$ requires fractions. Quadratic equations introduced square roots and complex numbers into the mix, and it was clear that an equation such as $x^3 - 2 = 0$ would require cube roots. Roots that are not whole numbers are known as radicals, and the goal was to find a formula that could be constructed from integers, radicals, and complex numbers that would give all solutions to the general cubic equation. Such a formula is referred to as "solution by radicals."

The first mathematician to make a dent in solving cubics by radicals was Scipione del Ferro, who late in the fifteenth century managed to find a formula that solved a restricted case of the general cubic, the case where $b = 0$. These "depressed cubics" have the form $ax^3 + cx + d = 0$, and del Ferro's mathematical fame would undoubtedly have increased significantly had the world learned of his advance. This, however, was an era in which Machiavelli was writing of the importance of subterfuge—and subterfuge, in Italian academe, was often how one survived.

Duels—of an intellectual nature—were one method by which up-and-comers of the time would obtain prestigious academic positions. A challenger would pose a list of questions, or mathematical problems, to an established academic, who would counter with a list of his own. After a predetermined amount of time, the results would be announced—as might be expected, to the victor belonged the spoils. The depressed cubic solution was del Ferro's ace in the hole—if challenged, he would present a list of depressed cubics to his challenger. As far as we know, del Ferro never had to play his trump card.[5]

A Duel of Wits with Equations As Weapons

Upon his death, del Ferro bequeathed the solution to his student Antonio Fior, a mathematician of less talent but greater ambition than his mentor. Del Ferro had kept the solution as a defense, but Fior decided to use it to make a name for himself, and issued a challenge to the well-known scholar Niccolo Fontana.

Fontana had been severely wounded as a child when a soldier slashed his face with a sword. This affected his speech, and led to his being given the nickname Tartaglia—the Stammerer—the name by which he is known today. When Fior presented his thirty-problem challenge to Tartaglia, Tartaglia countered with a list of thirty problems on a variety of mathematical topics—only to discover that Fior's list consisted of thirty depressed cubics.

It was a classic all-or-nothing situation—Tartaglia was either going to solve none of the problems, or all thirty, depending upon whether or not he could generate the solution to the depressed cubic. Tartaglia obtained the formula

$$\sqrt[3]{n/2 + \sqrt{m^3/27 + n^2/4}} - \sqrt[3]{-n/2 + \sqrt{m^3/27 + n^2/4}}$$

for the root of the depressed cubic $x^3 + mx = n$. As you can see, you are not likely to stumble upon this formula using hit-or-miss techniques.

I'll check out this formula with the depressed cubic $x^3 + 6x = 20$. The result is expressed as $x = \sqrt[3]{10 + \sqrt{108}} - \sqrt[3]{-10 + \sqrt{108}}$. Simplifying this expression is a good problem for an advanced high-school algebra student, but those who prefer modern technology can check with a pocket calculator that $x = 2$, which is indeed a solution of the equation.

There is a subtle deceptiveness to all mathematical textbooks that teachers realize but students generally don't; an awful lot of trial-and-error goes into establishing a result such as the solution of the depressed cubic, and most of that is error. We know that great composers such as Beethoven made sketchbooks of their ideas, and we can read them to discover some of the passages Beethoven considered using before arriving at the final version. Many mathematicians do the same—they keep records of their failed attempts, because sometimes what doesn't work for one problem might well solve another. However, these records generally don't make it into the archives, and as a result we have no idea how long it took del Ferro to discover his approach. Using modern notation, del Ferro's eventual successful solution isn't that difficult to follow.

We can divide our depressed cubic by the coefficient of x^3 to arrive at an equation that has the form

$$x^3 + Cx + D = 0$$

Instead of presenting the solution in the form it is usually given in a textbook, let's try to reconstruct what del Ferro did. Mathematicians often try different things in the hope of getting lucky, and so del Ferro tried assuming the solution had the form $x = s - t$. There's a valid reason for trying something like this, as using two variables (s and t) rather than one introduces an extra degree of freedom into the problem. This is a standard weapon in the mathematician's arsenal of problem-solving techniques, as the price one must pay for having to solve for additional variables may be more than offset by the ease of the solution. After making this substitution, the depressed cubic becomes

$$(s-t)^3 + C(s-t) + D =$$
$$(s^3 - 3s^2t + 3st^2 - t^3) + C(s-t) + D =$$
$$(s^3 - t^3 + D) - 3st(s-t) + C(s-t) =$$
$$(s^3 - t^3 + D) + (C - 3st)(s-t)$$

At this point, del Ferro undoubtedly realized he might have hit the jackpot. If he could find s and t such that $s^3 - t^3 + D = 0$ and $C - 3st = 0$, the last equation would become

$$0 + 0 \, (s-t) = 0$$

and $x = s \; t$ would be a root of the depressed cubic. So del Ferro was led to the system of two equations

$$3st = C$$
$$t^3 - s^3 = D$$

We are now left with the problem of finding s and t that satisfy the two equations; but here's how the bucket of water gets moved from under the table to the top of the table—these equations reduce to a quadratic! Solving $3st = C$ for t in terms of s yields $t = C/3s$, and substituting this into $t^3 - s^3 = D$ results in the equation

$$C^3/(27s^3) - s^3 = D$$

Multiplying through by s^3 and collecting all the terms on one side gives

$$s^6 + Ds^3 - (C^3/27) = 0$$

This equation is quadratic in s^3, for it can be written

$$(s^3)^2 + Ds^3 - (C^3/27) = 0$$

Using the quadratic formula, we obtain two possible solutions for s^3, but if the cube root of either is taken and t computed from the formula $t = C/3s$, the quantity $s - t$ will be the same and will solve the original depressed cubic.

It doesn't seem so difficult when it is neatly presented in a textbook, but when you have only a month and your future is at stake, it's a lot tougher. In a desperate race against time, an exhausted (no wonder) Tartaglia managed to find an ingenious geometrical approach to the problem, which yielded the solution just before the period allowed for the challenge expired. He solved all of Fior's problems, easily winning the contest. Tartaglia magnanimously did not require Fior to pay for losing the bet—in this case, he had bet thirty sumptuous feasts—but this may have been small consolation for Fior, who faded into obscurity as Tartaglia's fame increased.

Cardano and Ferrari—Scaling the Summit

One person to learn of Tartaglia's success was Girolamo Cardano, certainly one of the most unusual individuals ever to appear on the mathematical scene. Cardano was brilliant but bedeviled—afflicted with a number of infirmities, including hemorrhoids, ruptures, insomnia, and impotence. Compounding these physical problems was an assortment of psychological ones. He had acrophobia, an uncontrollable fear of mad dogs, and may not have been a masochist, but had formed the habit of inflicting physical pain upon himself because it was so pleasant when he stopped. We know all this because Cardano, who would have been a staple of late-night talk shows had such existed in the sixteenth century, wrote an extensive autobiography in which no details, no matter how intimate, seem to have been spared.

Cardano was fascinated by Tartaglia's victory, and wrote several letters imploring Tartaglia to tell him the secret of his success. Tartaglia responded with the sixteenth-century equivalent of "Sorry, my agent is working on a book deal," but Cardano persisted, and finally persuaded Tartaglia to leave his home in Brescia and visit Cardano in Milan. During this visit, Cardano managed to talk Tartaglia into revealing his secret—but in return, Tartaglia made Cardano take the following oath: "I swear to you by the Sacred Gospel, and on my faith as a gentleman, not only never to publish your discoveries, if you tell them to me, but I also promise and pledge my faith as a true Christian to put them down in cipher so that after my death no one shall be able to understand them."[6]

Like many of his contemporaries, Cardano placed great stock in dreams and omens, and was also a practicing astrologer. One night he dreamed of a beautiful woman in white, and he assiduously (and successfully) courted the first such woman who crossed his path, despite despairing of his chances; at the time he was poor as a church mouse. Soon after his meeting with Tartaglia, he heard a squawking magpie and believed that it presaged good fortune. When a young boy appeared at his doorstep looking for work, Cardano somehow saw this as the good fortune promised by the magpie and took him in. Maybe there was something to the squawking magpie theory, as the boy proved to have substantial mathematical ability. At first the boy, whose name was Ludovico Ferrari, was merely a servant in Cardano's household, but gradually Cardano taught him mathematics, and before Ferrari had reached his twentieth birthday, Cardano had passed on the secret of solving depressed cubics to him. The two mathematicians decided to tackle the problem of solving the general cubic.

Cardano and Ferrari achieved two major breakthroughs. The first was to find a transformation that reduced the general cubic equation to a depressed cubic, which Tartaglia's technique enabled them to solve. This transformation moves a different bucket of water from under the table to the top of the table.

Once again, by dividing by the coefficient of x^3, we can assume our general cubic equation has the form

$$x^3 + Bx^2 + Cx + D = 0$$

If we let $x = y - B/3$, this equation becomes

$$(y - B/3)^3 + B(y - B/3)^2 + C(y - B/3) + D = 0$$

Expanding the first two terms gives

$$(y^3 - By^2 + (B^2/3)y - (B^3/27)) + B(y^2 - (2B/3)y + (B^2/9)) + (y - B/3) + D = 0$$

It isn't necessary to completely simplify the left-hand side to note that there are only two terms involving y^2; the term $-By^2$ that occurs in the expansion of $(y - B/3)^3$, and the term By^2 that occurs in the expansion of $B(y - B/3)^2$; these terms cancel so the result is a depressed cubic in y, which del Ferro's technique enables us to solve for y. Then if $x = y - B/3$, x is a root of the original cubic.

That was the first breakthrough, but the second was even more exciting: Ferrari discovered a technique for transforming the general quartic equation (finding the roots of a polynomial of degree four) to a cubic, which they now knew how to solve. These were the most significant developments in algebra in millennia—but both advances ultimately rested on Tartaglia's solution to the depressed cubic, and Cardano's oath prevented them from publishing their results.

Several years later, Cardano and Ferrari traveled to Bologna, where they read the papers of Scipione del Ferro. These papers contained del Ferro's solution of the depressed cubic—which coincided with the solution that Tartaglia had found. Cardano and Ferrari managed to persuade themselves that since del Ferro had previously obtained the solution, using it would not break Cardano's pledge to Tartaglia.

Cardano published his classic work, *Ars Magna* ("the great art"), in 1545. Algebra was indeed Cardano's "great art"—though he was an accomplished physician (for his time) who had treated the pope, and though he wrote the first mathematical treatment of probability (Cardano was an inveterate gambler), his contributions to algebra are the ones for which he is best remembered. The description given earlier for the procedure used to solve the depressed cubic is taken from *Ars Magna*.

In *Ars Magna*, Cardano gave full credit to the giants on whose shoulders he stood. The preface to the chapter on the solution of the cubic began with the following paragraph: "Scipio Ferro of Bologna well-nigh thirty years ago discovered this rule and handed it on to Antonio Maria Fior of Venice, whose contest with Niccolo Tartaglia of Brescia gave Niccolo occasion to discover it. He gave it to me in response to my entreaties, though withholding the demonstration. Armed with this assistance, I sought out its demonstration in [various] forms. This was very difficult."[7]

Tartaglia did not take this revelation well, accusing Cardano of violating his sacred oath. Cardano did not reply to these accusations, but Ferrari, who was known as something of a hothead, did. This culminated in a challenge match between Tartaglia and Ferrari—but Ferrari had the home-field advantage and emerged victorious. Tartaglia blamed his defeat on the vigor with which the onlookers supported the home favorite (there's something quaintly charming about a citizenry that will riot in response to a contest of the intellect rather than, as is the case nowadays, in response to the results of a soccer match, but possibly there wasn't a whole lot to root for back in the sixteenth century). Ferrari, naturally, felt that his own brilliance was responsible for his victory. At this point, the story of the search for a solution by radicals to polynomial equations pauses for two centuries, awaiting the arrival of the final characters in this drama.

The travails suffered by many of the major characters in this drama are the stuff of which miniseries are made. Cardano's wife died young, his elder son Giambattista was executed for murder, and his other son was imprisoned for criminal activities. Cardano himself was thrown in jail for heresy (not a good era in which to be a heretic), but he was later pardoned. Cardano's epitaph might well be the last line of his *Ars Magna*: "Written in five years, it may last thousands."[8] Ludovico Ferrari died from poison, which many historians believe was administered by his sister.

The Insolubility of the Quintic

The general cubic had been solved by reducing it to a depressed cubic, and the quartic had been solved by reducing it to a cubic—but the solutions to each polynomial of higher degree were becoming ever more involved and complicated. It appeared that the future of solving the general quintic—the polynomial of degree five—was going to follow the same path: find the transformation that reduced it to a quartic, and then use Ferrari's formula. This seemed a rather dreary prospect. Perhaps that's why more than two centuries passed, and though mathematics made considerable advances, most were in calculus and related areas. Trying to find the general solution to the quintic was no longer a top priority of the mathematical community—calculus was newer and a whole lot sexier.

As sometimes happens in both mathematics and science, the tools available to the community are simply inadequate for solving certain problems, and the mathematical or scientific community hits the wall. New and different techniques are required—although often the community simply doesn't realize it until those techniques actually make an appearance. Such was the case with the solution of the quintic. The resolution of this problem did not occur until the turn of the nineteenth century, when three brilliant mathematicians broke new ground with a totally different approach, one which was to forever alter the direction of mathematics.

Paolo Ruffini

For nearly 250 years after Cardano and Ferrari had solved the quartic, mathematicians had tried to crack the mystery of the quintic. Some of the great names of mathematics foundered on the shoals of this problem, including Leonhard Euler and Joseph-Louis Lagrange. The latter published a famous paper, *Reflections on the Resolution of Algebraic Equations,* in which he stated that he planned to return to the solution of the quintic, which he obviously hoped to solve by radicals.

Paolo Ruffini was the first mathematician to suggest that the quintic could not be solved by radicals, and he offered a proof of it in *General Theory of Equations in Which It Is Shown That the Algebraic Solution of the General Equation of Degree Greater Than Four Is Impossible.* In it, he states, "The algebraic solution of general equations of degree greater than four is always impossible. Behold a very important theorem which I believe I am able to assert (if I do not err): to present the proof of it is the main reason for publishing this volume. The immortal Lagrange, with his sublime reflections, has provided the basis of my proof."[9]

Unfortunately, that introduction turned out to be prescient—there was a gap in his proof. However, not only had Ruffini glimpsed the truth, he had realized that the path to the solution led through an analysis of what happened to equations when the roots of a polynomial were permuted. Even though he did not formalize the idea of a permutation group, he proved many of the initial basic results in the theory.

Ruffini was yet another mathematician to be dogged by bad luck. He never really received credit for his work—at least in his lifetime. The only top mathematician to give him the respect he deserved was Augustin-Louis Cauchy, but when his paper was examined by leading French and English mathematicians, the reviews were neutral (the English) to unfavorable (the French). Ruffini was never notified that his proof contained a gap—had a leading mathematician done so, he would have had a shot at patching the proof. Usually, the person most familiar with a flawed proof has the best chance of fixing it—but Ruffini was never given the chance.

Groups in General—Permutation Groups in Particular

One of the most important accomplishments of mathematics is that it has shown that apparently dissimilar structures possess many important common attributes. These attributes can be codified into a set of axioms, and conclusions derived for all structures that satisfy those axioms. One of the most important such objects is called a group.

To motivate the definition of a group, consider the set of all nonzero real numbers. The product of any two nonzero real numbers x and y is a nonzero real number xy; this product satisfies the *associative* law: $x(yz) = (xy)z$. The number 1 has the property that for any nonzero real number, $1x = x1 = x$. Finally, each nonzero real number has a multiplicative inverse x^{-1} that satisfies $xx^{-1} = x^{-1}x = 1$. These are the key properties used to define a group G, which is a collection of elements and a way of combining two of those elements g and h into an element gh in G. This way of combining elements is usually referred to as multiplication, and the element gh that results is called a product, although as we shall see there are many groups in which "multiplication" bears no resemblance to arithmetic. The multiplication must satisfy the associative law: $a(bc) = (ab)c$ for any three elements a, b, and c of the group. The group must contain an identity element, which could be denoted by 1, which satisfies $g1 = 1g = g$ for any member g of the group. Finally, each member g of the group must have a multiplicative inverse g^{-1}, which satisfies $gg^{-1} = g^{-1}g = 1$.

An interesting example of a group, which has an important and surprising connection to the problem of solving the quintic, is found by examining what happens when we shuffle a deck of cards. It is possible to completely describe a shuffle by thinking of where cards end up relative to where they start. For instance, in a perfect shuffle, the top twenty-six cards are placed in the left hand and the bottom twenty-six cards in the right. The mechanics of the classic "waterfall" shuffle releases the bottom card from the right hand, then the bottom card from the left, then the next-to-bottom card from the right hand, and so on, alternating cards from each hand. We could describe the perfect shuffle by means of the following diagram, which describes where a card starts in the deck and where it ends up; the top card in the deck is in position 1, and the bottom card in position 52.

Starting Position	1	2	3	...	24	25	26	27	28	29	...	50	51	52
Ending Position	1	3	5	...	47	49	51	2	4	6	...	48	50	52

We could produce a shorthand for this using algebraic notation.

Starting Position (x)   Ending Position

$$1 \leq x \leq 26 \qquad 2x - 1$$
$$27 \leq x \leq 52 \qquad 2x - 52$$

The set of all shuffles of a deck of cards forms a group. The product gh of two shuffles g and h is the rearrangement that results from first performing shuffle g, then shuffle h. The identity element of this group is the shuffle that doesn't change the position of any card—the "phantom shuffle" that is sometimes performed by magicians or cardsharps. The inverse of any shuffle is the shuffle that restores the cards to their original position. For instance, we can use the above diagram to get a look at a portion of the inverse to the perfect shuffle.

Starting Position	1	2	3	4	...	49	50	51	52
Ending Position	1	27	2	28	...	25	51	26	52

Again, using algebraic notation.

Starting Position (x)	Ending Position
x is odd	$(x+1)/2$
x is even	$26+x/2$

To see that this is indeed the inverse of the perfect shuffle, notice that if a card starts out in position x, where $1 \leq x \leq 26$, the perfect shuffle puts it in position $2x-1$ (an odd number), so the inverse puts it in position $((2x-1)+1)/2=x$—back where it started. If a card starts in position x, where $27 \leq x \leq 52$, the perfect shuffle puts it in position $2x-52$ (an even number), so the inverse puts it in position $26+(2x-52)/2=x$—again, back where it started. Similarly, one can show that if one performs the inverse first and follows it with the perfect shuffle, every card returns to its original position. Although it is not germane to the quintic problem, performing eight perfect shuffles of a deck of fifty-two cards restores the deck to its original order—if g denotes the perfect shuffle, this is written $g^8=1$, and mathematicians say that g is an element of order 8. Showing that shuffling satisfies the associative law is not difficult—but it isn't especially interesting, so I'll skip the demonstration.

Notice that the perfect shuffle—and its inverse—leave the top card of the deck unchanged. If we were to consider all the shuffles that leave the top card of the deck unchanged, we would discover that they also form a group—the product of any two such shuffles leaves the top card unchanged, and the inverse of such a shuffle also leaves the top card unchanged. A subset of a group that is itself a group is called a subgroup.

One way in which the group of all shuffles differs from the group of nonzero real numbers is that the latter group is commutative—no matter which order you multiply two numbers, the result is the same: for example, $3 \times 5 = 5 \times 3$. The same cannot be said of shuffles. If shuffle g simply flip-flops the top two cards (and leaves the other cards in the same position), and if shuffle h just flip-flops the second and third card, let's follow what happens to the third card in the deck. If we perform g first, the third card stays where it is, but then migrates to position 2 after we then perform h. If we perform h first, the third card initially moves to position 2, and then g moves it to position 1. So performing the shuffles in different orders produce different results—the order of shuffling (multiplication in this group) does make a difference.

Although a standard deck of cards contains fifty-two cards, one could obviously shuffle a deck of any number of cards. The group of all possible shuffles of a deck of n cards is known as the symmetric group S_n. The structure of S_n—that is, the number and characteristics of its subgroups—becomes more complex the higher the value of n, and this is the key fact that determines why the quintic has no solution in terms of radicals.

Niels Henrik Abel (1802–1829)

Niels Henrik Abel was born into a large, and poor, Norwegian family. At the age of sixteen, he embarked upon a program of reading the great works of mathematics; but when he was eighteen, his father died. Abel, though not in good health himself, assumed the responsibility for taking care of his family. Despite these obligations, he decided to attack the quintic, and initially thought he had obtained a solution in the manner of Cardano and Ferrari. After realizing that his proof was in error, he came to precisely the opposite conclusion: it was impossible to find an algebraic expression for the roots of the general quintic. Working along the same general lines as Ruffini, but avoiding the proof pitfalls that had plagued the Italian mathematician, Abel was able to show that the general quintic could not be solved by radicals, bringing to an end a quest that had started more than three millennia earlier in Egypt.

After publishing a memoir outlining his proof, Abel went to Berlin, where he began publishing his results on a variety of topics in the newly launched *Crelle's Journal*. These results were favorably viewed by German mathematicians, and Abel then traveled to Paris, where he hoped to obtain recognition from the leading French mathematicians.

However, France was a hotbed of mathematical activity, and Abel wrote to a friend, "Every beginner has a great deal of difficulty getting noticed here."[10] Discouraged and weakened by tuberculosis, Abel returned home, where he died at the tragically young age of twenty-seven. Unbeknownst to Abel, his papers had been generating increasing excitement in the mathematical community, and two days after his death a letter arrived bearing an offer of an academic position in Berlin.

Évariste Galois

The third major player in the solution of the quintic also suffered similarly from bad luck. Évariste Galois was born nine years later than Abel in a suburb of Paris. The son of a mayor, he did not exhibit any exceptional ability in school—but by age sixteen he realized that despite the judgments of his teachers, he possessed considerable mathematical talents. He applied to the École Polytechnique, a school which had been attended by many celebrated mathematicians, but his mediocre performance in school prevented him from being accepted. He wrote a paper and presented it to the academy at age seventeen—but Augustin-Louis Cauchy, one of the leading mathematicians of the era, lost it. He submitted another paper to the academy shortly thereafter—but Joseph Fourier, the secretary of the academy, died soon after the receipt of the paper and it, too, was lost. Jonathan Swift once remarked that one could recognize genius by the fact that the dunces would conspire against them; Galois seems to have been particularly unfortunate in that geniuses conspired against him, albeit inadvertently.

Frustrated by all this incompetence, Galois sought an outlet in the politics of the times, and joined the National Guard. An active revolutionary, in 1831, he proposed a toast at a banquet that was viewed as a threat against King Louis Philippe. This declaration was followed by a mistake that was to prove fatal—he became involved with a young lady whose other lover challenged Galois to a duel. Fearing the worst, Galois spent the night before the duel jotting down his mathematical notes, entrusting them to a friend who would endeavor to have them published. The duel took place the next day, and Galois died from his wounds a day later. He was barely twenty years old.

Although Abel was the first to show the insolvability of the quintic, Galois discovered a far more general approach to the problem that was to be of great significance. Galois was the first to formalize the mathematical concept of a group, which is one of the central ideas in modern algebra. The connection between groups, polynomial equations, and fields is one of the primary themes of the branch of mathematics known as Galois theory. Galois theory not only explains why there is no general solution to the quintic, it also explains precisely why polynomials of lower degree have solutions. Remarkably, Galois theory also provides clear explanations of three compass-and-straightedge impossibilities we have previously examined: why the cube cannot be duplicated, why the angle cannot be trisected, and why only certain regular polygons are constructable.

Galois Groups

When I first learned the quadratic formula in high school, my algebra teacher mentioned that there were such formulas for polynomials of degree three and degree four, but no such formula existed for polynomials of degree five. At the time, I didn't completely understand what the teacher meant, and interpreted his remark to mean that mathematicians simply hadn't discovered the formula yet. It wasn't until later that I realized that although there were formulas that did give the roots of fifth-degree polynomials, those formulas used expressions other than radicals—if the "language" for describing solutions consisted simply of whole numbers, radicals, and algebraic expressions involving them, then that language simply doesn't have a means of expressing the roots of all fifth-degree polynomials. One of my goals as a student was to find out why this was so—but in order to fully understand it, one must learn Galois theory. In order to understand Galois theory, it is necessary first to take an introductory course in abstract algebra, which usually comes about the third year of college.

Nonetheless, it is possible to understand some of the basic ideas surrounding the theory. Using the quadratic formula, the polynomial $x^2 - 6x + 4$ has two roots: $A = 3 + \sqrt{5}$ and $B = 3 - \sqrt{5}$. These roots satisfy two basic algebraic equations: $A + B = 6$ and $AB = 4$. Admittedly, they satisfy a whole bunch more, such as $5(A + B) - 3(AB)^3 = 5 \times 6 - 3 \times 64 = -162$, but this equation was obviously constructed from the other two. They also satisfy $A - B = 2\sqrt{5}$, but this equation is qualitatively different from the first two: the only numbers that appear in the first two equations are rational numbers, whereas the last equation contains an irrational number. Notice also that if we tried something like $A + 2B$, we would get the irrational number $9 - \sqrt{5}$, so the equations that can be constructed from A and B that involve only rational numbers are definitely limited.

Look once again at the two equations $A + B = 6$ and $AB = 4$, but instead of writing them in this form, write them in the form $\square + \triangle = 6$ and $\square \triangle = 4$, where the plan is to look at the various possible ways of inserting the two roots A and B into the $\square$ and $\triangle$ locations in order to get a true statement. There are two ways that this can be done. One is the original way we obtained these equations—insert A into the $\square$ and B into the $\triangle$, which gives the original two (true) statements $A+B=6$ and $AB=4$. If a deck of two cards is shuffled, where A is initially on top and B is initially on the bottom, then $\square$ represents the letter that ends on top and $\triangle$ the letter that ends on the bottom after the shuffle. The substitution A for $\square$ and B for $\triangle$ corresponds to the phantom shuffle. The only other shuffle of two cards has A ending on the bottom and B on top, so when B is substituted for $\square$ and A for $\triangle$, the resulting statements $B+A=6$ and $BA=4$ are still true. The Galois group of a polynomial consists of all those shuffles that result in all the algebraic equations with rational numbers being true statements. So the Galois group of the polynomial x^2-6x+4 consists of the two shuffles (phantom and switch top-two cards) that comprise S_2.

It is not always the case that both shuffles in S_2 are in the Galois group of the polynomial. To see such a case, consider the polynomial x^2-2x-3, whose two roots A and B are 3 and -1. The two roots satisfy $A+2B=1$, so examine the algebraic equation with rational coefficients $\square+2\triangle=1$. If A and B are switched in the left side of the equation, the resulting equation is $B+2A=1$, which is not a true statement, as the sum $B+2A$ actually equals 5. For this polynomial, the only shuffle that generates true statements from the original equations is the identity element (the phantom shuffle), so in this case the Galois group of x^2-2x-3 consists of just the phantom shuffle.

There is a famous quote from the American astronomer Nathaniel Bowditch, who translated Laplace's *Celestial Mechanics* into English. Bowditch remarked, "I never come across one of Laplace's 'Thus it plainly appears' without feeling sure that I have hours of hard work before me to fill up the chasm and find out and show how it plainly appears."[11] The same is generally true for the statement "It can be shown," so I am loathe to include it unless I absolutely must—but here I absolutely must. It can be shown that a polynomial has roots that can be expressed in terms of radicals only when its Galois group has a particular structure in terms of its subgroups. This structure is known as solvability; it is quite technical to describe, but the name is clearly motivated by the problem of solving the problem of finding the roots of a polynomial by radicals. The Galois group of the polynomial x^5-x-1 can be shown (oops, I did it again) not to be solvable, and so the roots of that polynomial cannot be found by radicals.

Later Developments

The insolvability of the quintic proved to be a significant moment in the development of mathematics. It is not possible to say with certainty what would have happened had quintics, and polynomials of higher degree, proven to have solutions by radicals, but one can say with some assurance that mathematics is a lot more interesting a subject because the quintic does not have such a solution.

Mathematics is a language used to describe a variety of phenomena—but a language needs words. Some of the most important words in the mathematical language are functions. Functions, such as powers or roots, can be combined in two basic ways—algebraically (using addition, subtraction, multiplication, and division), and compositionally (one after the other, like successive shuffles—one can square a number and then take its cube root). The insolvability of the quintic amounts to a declaration that the vocabulary of functions that can be constructed with powers and roots is inadequate to describe the solutions of a certain equation. This naturally stimulated a search for other functions that could be used to describe these solutions.

Where do functions come from? Often they arise from need. The trigonometric functions are used for expressing quantities determined by angles, as well as in describing periodic phenomena, and the exponential and logarithmic functions are used for describing growth and decay processes. Many functions arise as solutions to important equations (usually differential equations) that occur in science and engineering. For example, Bessel functions (named after the nineteenth-century mathematician and physicist William Bessel, who was the first to calculate the distance to a star) occur as solutions to the problem of how a membrane such as a drum vibrates when it is struck, or how heat is conducted in a cylindrical bar.

In 1872, the German mathematician Felix Klein was able to find a general solution for the quintic in terms of hypergeometric functions, a class of functions that occur as a solution to the hypergeometric differential equation.[12] In 1911, Arthur Coble solved the sextic, the general polynomial of degree six, in terms of Kampé de Fériet functions—a class of functions of which I had never heard and I doubt that 99 percent of living mathematicians have, either. The trend appears bleak—it looks as if the general solution to polynomials of ever-higher degree, if such solutions can be found, will be in terms of ever-more obscure classes of functions. Functions are indeed like words: their utility depends largely on the frequency with which they are used, and functions (or words) that are so specialized that only a few know them have limited value.

The solving of equations is central not only to mathematics, but to the sciences and engineering. Mathematicians may be interested to know that the solution to a particular equation exists, but to build something it is necessary to know what that solution is—and to know it to three, five, or eight decimal places. Numerical analysis is not, as its name would suggest, the analysis of numbers; it is the branch of mathematics that deals with finding approximate solutions to equations—to an accuracy of three, five, eight (or whatever) decimal places. Knowing that it may not be possible to find an exact formula for the solution to an equation, yet realizing that to build something may require an accurate approximation to that solution, impelled mathematicians to devise techniques for finding these approximate solutions and, equally important, knowing how accurate these solutions are. An inexpensive pocket calculator will give the cube root of 4 as 1.587401052; but if this number is cubed, the answer will not be 4—although it will be very close to it. The cube root of 4 as given by the calculator is accurate to nine decimal places—good enough for building all mechanical devices and many electronic ones. From a practical standpoint, numerical analysis can generally determine the roots of polynomials with sufficient accuracy to build anything whose construction depends upon knowing those roots.

At the moment, though, the quest for solutions of polynomials is going in new directions. Just as the search for the roots of polynomials took an abrupt turn at the dawn of the nineteenth century and brought group theory into the picture, relatively new branches of mathematics are currently being brought to bear on the problem. Many of the most widely studied groups are connected with symmetries of objects. For instance, we have looked at S_3 as the set of all shuffles of a three-card deck. However, if one imagines an equilateral triangle with vertices A, B, and C, initially starting with A as the top vertex and B and C as the bottom left and bottom right vertices, the triangle can be rotated or reflected so that the new position of the triangle corresponds to one of the shuffles.

Triangle	1		2		3		4		5		6	
Top Vertex	A		C		B		A		C		B	
Bottom	B	C	A	B	C	A	C	B	B	A	A	C

We can actually see how the group structure arises in this example—there are two fundamentally different operations from which the others are constructed. These are a counterclockwise rotation of 120 degrees, which we could denote by R. Triangle 2 is obtained from triangle 1 by doing R. The other basic operation is to leave the top vertex unchanged but flip the bottom two; we denote this by F. Triangle 4 is obtained from triangle 1 by doing F. Similarly, triangle 3 is obtained from triangle 1 by performing R twice; this operation is denoted RR, or R^2. Triangle 5 is obtained from triangle 1 by performing R first, then F, denoted RF; triangle 6 is likewise obtained by performing F first and then R— or FR.

This is essentially the same group as the shuffles of a three-card deck—one can identify R with the shuffle that simply puts the top card on the bottom of the deck, and F with the shuffle that leaves the top card alone but switches the position of the second and third cards. This process of identifying two apparently different groups with each other is known as isomorphism—a process that enables mathematicians to translate truths known about one object to truths known about the other. The proof that the general quintic has no solution involves a group isomorphic to the group of symmetries of the regular icosahedron—the regular platonic solid with twenty faces, all of which are equilateral triangles. Mathematicians nowadays are looking to geometry in the hope that they can discover things that will translate into problems involving roots of polynomials.

The French politician Georges Clemenceau once said that war was too important to be left to the generals. Similarly, group theory was too important to be left to the mathematicians. Group theory is employed extensively in the sciences, because group theory is the language of symmetries, and science has discovered that symmetry plays a fundamental role in many of its laws. I'm not sure whether anyone has written *Group Theory for Anthropologists* or *Group Theory for Zoologists,* but there are books with similar titles written for biochemists, chemists, engineers—probably the greater part of the alphabet, and I'd be willing to bet that every letter of the alphabet is represented when it comes to describing types of groups (we already observed that the letter *s* is used for "solvable group"). Noticing patterns, and missing elements of patterns, is often the key to important discoveries, and group theory provides an organizing framework that often points the way to the missing element.

The story of the search for solutions by radicals to polynomial equations did not end with the discovery that one could not find formulas for the quintic; rather, it branched off to generate useful and exciting results that even Cardano and Ferrari, who scaled the summit of what could be done in this area, would undoubtedly have found every bit as enchanting as those revealed in Cardano's *Ars Magna.*

NOTES

1. See http://history1900s.about.com/od/1950s/a/hopediamond.htm.
2. It would be more accurate to say that polynomials are the only everywhere differentiable functions we can calculate. For example, the function $f(x)$, beloved of

analysts, defined by $f(x)=0$ if x is rational and $f(x)=1$ if x is irrational, can be calculated for every valuable of the variable. This function is highly artificial, as it never shows up in any process related to the real world.

3. A. B. Chace, L. S. Bull, H. P. Manning, and R. C. Archibald, *The Rhind Mathematical Papyrus* (Oberlin, Ohio: Mathematical Association of America, 1927–29). A good description of this method, known as the method of false position, can be found at http://www-groups.dcs.st-and.ac.uk/~history/HistTopics/Egyptian_papyri.html.

4. An example of solving an equation by completing the square is

$$x^2 - 4x - 5 = 0$$
$$x^2 - 4x = 5$$
$$x^2 - 4x + 4 = 5 + 4 \text{ (the "completing-the-square" step)}$$
$$(x-2)^2 = 9$$
$$x - 2 = 3 \text{ or } -3$$
$$x = 5 \text{ or } x = -1$$

5. W. Dunham, *Journey Through Genius* (New York: John Wiley & Sons, 1990).

6. As quoted at http://www-history.mcs.st-andrews.ac.uk/Biographies/Cardan.html.

7. G. Cardano, *Ars Magna* (Basel, 1545).

8. Ibid.

9. As quoted at http://www-groups.dcs.st-and.ac.uk/~history/Biographies/Ruffini.html.

10. Carl B. Boyer, *A History of Mathematics* (New York: John Wiley & Sons, 1991), p. 523.

11. Quoted in F. Cajori, *The Teaching and History of Mathematics in the United States* (Whitefish, MT: Kessinger Publishing, 2007).

12. The geometric series with ratio r is the infinite sum $1 + r + r^2 + r^3 + \ldots$; hypergeometric series generalize this series. A more extensive discussion of this topic can be found at http://en.wikipedia.org/wiki/Hypergeometric_functions, but unless you plan on a career that requires a substantial knowledge of advanced mathematics, you can skip this discussion.

6 Never the Twain Shall Meet

It Takes an Adult

There are some pleasures that even three-year-olds can enjoy, such as ice cream and warm sunshine on your face on a nice spring day—but there are some pleasures that are reserved for adults. Intelligent conversation. Vegetables. Geometry.

Believe me, I didn't wake up at age eighteen and say, "Math looks interesting—maybe I should major in it." I liked math ever since we first started counting things back in kindergarten, or even before. Well, I liked math until I hit geometry, and then there was a miserable year when I struggled to get Bs because I kept not seeing how to prove some things and skipping steps in things I could see how to prove. Advanced algebra and trig rekindled my enthusiasm, and when I got to analytic geometry and calculus I was back on the math track—partially because these two subjects almost completely eliminate the necessity for knowing anything but basic stuff in geometry.

I can't remember what bothered me most about geometry, but I do remember that indirect proofs were near the top of the list. An indirect proof is one in which you assume the negation of a conclusion, show that this leads to a contradiction, and consequently the only option left is that the sought-after conclusion must be correct. A large number of indirect proofs in geometry are the result of Euclid's infamous fifth postulate—the parallel postulate.

Noncontroversial Geometry

Noncontroversial geometry is everything up to, but not including, the parallel postulate. It includes basic objects that we can't really define but everybody knows what they are, some definitions involving basic objects, some obvious arithmetic and geometric facts, and the four postulates that precede the parallel postulate.

Basic objects are things like points. Euclid defined a point as that which has no part.[1] Works for me—I'm not philosopher enough to say exactly what these abstract constructs are, but I (and you) know what Euclid was getting at, so we can move on. Obvious arithmetic facts were such statements as equals added to equals are equal. Euclid's one obvious geometric fact was that things that coincided with each other were equal—if line segments AB and CD can both be positioned to coincide, then $AB = CD$.

We come now to the four noncontroversial postulates. I'll assume that line segments have endpoints, but straight lines don't. Using this terminology, the postulates are

> Postulate 1: Any two points can be connected by a unique line segment.
> Postulate 2: Any line segment can be extended to a straight line.
> Postulate 3: There is a unique circle with given center and radius.
> Postulate 4: All right angles are equal.[2]

There is an amazing amount of geometry that can be done using only those four postulates—but that doesn't concern us here.

The Parallel Postulate

Euclid's initial version of the parallel postulate was, to say the least, unwieldy.

Postulate 5 (Euclid): If a straight line falling on two straight lines makes the interior angles on the same side less than two right angles, the two straight lines, if extended indefinitely, meet on that side on which the angles are less than two right angles.[3]

To understand what is happening here, think of a triangle with all its sides extended indefinitely. Look at the side you consider to be the base of the triangle. The "interior angles" referred to above are the angles the base makes with the other two sides; the sum of those two angles is less than the 180 degrees, that is, the sum of the two right angles. If one changed the orientation of the sides by a sufficient amount that they intersected on the *other* side of the base, the "interior angles" here would again sum to less than 180 degrees. So what happens when the interior angles sum to precisely 180 degrees? The two other lines don't meet on either side of the base, so they either have to meet on the base (but that would happen only if the base coincided with the sides) or not meet.

As you can see, this formulation of the parallel postulate is not easy to work with, and even back in ancient Greece suggestions were made to revise it. It was Proclus who suggested a version that we frequently use today (two parallel lines are everywhere the same distance apart), but it was the Scottish mathematician John Playfair who gets the credit, as he wrote a very popular geometry text at the turn of the nineteenth century incorporating it. Then as now, credit accrues to the individual with the best public relations department.

Postulate 5 (Playfair's Axiom): Through each point not on a given line, only one line can be drawn parallel to the given line.[4]

This was the form in which I learned the parallel postulate. It has two obvious advantages. The first is that it is much easier to understand, visualize, and use than Euclid's original formulation. The second advantage is more subtle—it leads one to ask the question, is it possible to create geometries in which more than one line can be drawn parallel to the given line?

Certainly, such a geometry cannot exist on the plane, as that's the habitat of Euclidean *plane* geometry with the five postulates. However, if we move into Euclidean three-dimensional space, we can have infinitely many lines through a given point parallel to a given line—parallel, that is, in the sense that both lines, when extended, do not meet. Simply take a line and a plane parallel to that line but not containing it. If one fixes a point in that plane, any line through that point will obviously not meet the given line—although all but one of these are called skew lines in modern terminology (there is one that is genuinely parallel to the given line because it lies in a plane with the given line).

Girolamo Saccheri

Not all Italian mathematicians were as colorful as Tartaglia, Cardano, and Ferrari. Girolamo Saccheri was ordained a Jesuit priest and taught philosophy and theology at the University of Pavia. He also held the chair of the mathematics department there, leading me to wonder if that situation was something like what happened during the Great Mathematics Teacher Drought of the 1970s, where the paucity of math teachers at the junior and senior high school level sometimes resulted in shop or PE teachers becoming algebra instructors. A good friend of mine had majored in political science while she was in college. When she became a middle-school teacher (the West Coast equivalent of junior high) in the 1970s, someone was needed to fill an algebra staffing gap; she did so and had a satisfying and successful career as an algebra teacher.

Anyway, not much was heard from Saccheri until 1733, when his bombshell *Euclides ab Omni Naevo Vindicatus* (variously translated, I'll go with "Euclid Freed from Every Flaw") was published. It was more of a time bomb, not being recognized for its value until substantially later. In it, Saccheri was to make the first important moves toward the development of non-Euclidean geometry.

Saccheri did what others before him attempted to do—prove the parallel postulate from the other four postulates. He started with a line segment (the base) on which he constructed two line segments of equal length (the sides), each making a right angle with the base. He then connected the endpoints of the two line segments (the top), making a figure that is now known as a Saccheri quadrilateral—and which you, when you do this, will immediately recognize as a rectangle.

However, you know it's a rectangle because each point of the top is the same distance from the base (the sides of equal length make it so), and you have accepted Proclus's version of the parallel postulate. Saccheri didn't assume the parallel postulate. By using the other postulates, he was able to show quite easily that the vertex angles, which are the two angles made by the top with the sides, had to be equal. There were then three possibilities: the vertex angles could be right angles (which would then demonstrate that the parallel postulate could be proved from the other four), the vertex angles could be obtuse (greater than 90 degrees), or the vertex angles could be acute (less than 90 degrees).

Saccheri first developed an indirect proof in which he showed that the hypothesis that the vertex angles were obtuse led to a contradiction. He then attempted to show that the hypothesis that the vertex angles were acute also led to a contradiction—but after much work was unable to do so without fudging the proof by assuming that lines that met at a point at infinite distance (this is called "a point at infinity") actually met at a point on the line. At this juncture, Saccheri had two choices—go with the fudged proof in order to show the result in which he had an emotional investment, or admit that he was unable to show that the hypothesis that the vertex angles were acute led to a contradiction. In retrospect, had he chosen the second option, he could possibly have advanced the discovery of non-Euclidean geometries by decades—but he went with the first.

Saccheri also was the first to realize an important property of non-Euclidean geometries: the assumption that the vertex angles were acute led to the conclusion that the sum of the angles in a triangle must be less than 180 degrees. Most investigations of whether the universe is Euclidean or non-Euclidean involve measuring the angles of a triangle—the larger the triangle the better—to see if this measurement will reveal the underlying geometry of the universe. A triangle the sum of whose angles is less than 180 degrees, and such that the result is outside the range for experimental error, would unquestionably show that the universe was non-Euclidean. However, a triangle the sum of whose angles is close to 180 degrees would only provide confirming evidence that the universe was Euclidean, and would not constitute a definitive result.

Another Visit from the Dancing Angels

Saccheri published his results in 1733. Some thirty years later, the German mathematician Johann Lambert, a colleague of Leonhard Euler and Joseph Lagrange, took another shot at the problem using a very similar approach. Instead of using Saccheri quadrilaterals (two right angles with two equal sides), he looked at a quadrilateral with three right angles and deduced conclusions about the fourth angle using postulates one through four. Like Saccheri, he disposed of the possibility that the fourth angle could be obtuse, but unlike Saccheri, he recognized that no contradiction could be obtained if one assumed that the fourth angle were acute. Under the assumption that the fourth angle was acute, Lambert managed to prove several important propositions about models for non-Euclidean geometry—much as George Seligman, my undergraduate algebra teacher, had managed to prove results about algebras of dimension 16. However, Lambert did not construct models for non-Euclidean geometries, so at the time of his death it wasn't clear whether angels could dance on the head of this particular pin. Lambert would be more fortunate than Seligman, as help was on its way—but the final verdict would not be in for nearly a century.

An Unpublished Symphony from the Mozart of Mathematics

At the turn of the nineteenth century, three mathematicians were to travel essentially the same path toward the construction of non-Euclidean geometries—and they all did it in essentially the same fashion, by substituting an alternative for Playfair's Axiom. Each of the three worked with "Through each point not on a given line there exists more than one parallel to the given line," and each deduced much the same conclusions—although history gives the lion's share of the credit to Nikolai Ivanovich Lobachevsky and János Bolyai.

Although Gauss was undoubtedly the first to reach the conclusion that a consistent geometry was possible using the above alternative to Playfair's Axiom, Gauss lived in a different era—and played mathematics by a different set of rules than those commonly used today. Gauss's unofficial motto was *Pauca, Sed Matura*—which translates from the Latin as "Few, but ripe," and expresses his attitude toward publishing. Gauss did not publish anything until he was convinced that doing so would add to his prestige (which, considering his prestige, meant he would publish only the crème de la crème), and also that the result had been polished to a fare-thee-well. Of course, like any mathematician, he certainly did not burn his papers, and he was willing to communicate his results privately. One day, he received a visit from Carl Jacobi, at the time generally regarded as the second-best mathematician in Europe. Jacobi wanted to discuss a result he had obtained, but Gauss extracted some papers from a drawer to show Jacobi he had already obtained the result. A disgusted Jacobi remarked, "It is a pity that you did not publish this result, since you have published so many poorer papers."[5]

Newton had probably set the gold standard for recalcitrance when it came to publishing. He stuck his work on gravity in a drawer—probably as the result of a vicious academic dispute with Robert Hooke over the nature of light. Some years later, the astronomer Edmond Halley (of Halley's comet fame) came to visit Newton, and inquired of him what would be the motion of a body under an inverse square law of gravitational attraction. Newton astounded Halley by telling him that he had calculated it to be an ellipse, and Halley was so impressed that he underwrote the cost of publishing Newton's *Principia*—which Newton had difficulty finding when Halley visited him, because he wasn't sure where he had hidden it. When Newton didn't avoid publication, he published anonymously—but his solution to a problem posed by Johann Bernoulli was so elegant that even though the solution was anonymous, Bernoulli knew it was Newton's, declaring that he knew the lion by his claw.

Publication is a very different matter nowadays. With rare exceptions (such as when Andrew Wiles announced a solution to Fermat's last theorem), mathematicians generally publish, or try to publish, what they've got—even if it isn't a polished solution to a problem, or even a complete one.

There are good reasons for this. Young mathematicians, especially at prestigious universities, are well aware of the adage "publish or perish." The final tenure decision on assistant professors generally occurs no later than six years after the initial hiring, and no matter how good a teacher you are, at a top-ranked university, you'd better have something to show, publication wise, for those six years—or you're going to be looking for another job. As a result, the pressure to publish—even prematurely—is enormous. Additionally, even for the tenured, getting something out there is important because (1) it helps to make a contribution, and (2) by doing so, you may supply the critical piece of the puzzle that can turn an unproven result, or Someone Else's theorem, into Yours and Someone Else's theorem. I know, because I read a paper by the esteemed Czechoslovokian mathematician Vlastimil Pták, had one of the few really good ideas I've had,[6] and wrote up a short paper that appeared in the *Proceedings of the American Mathematical Society*. The best result in this paper was to become known as the Pták-Stein theorem (as far as I know, the only thing that's named after me)—and nine months later a paper appeared elsewhere with the exact same result. As Tom Lehrer put it in his hilarious song "Nikolai Ivanovich Lobachevsky,"

> And then I write
> By morning, night,
> And afternoon,
> And pretty soon
> My name in Dnepropetrovsk is cursed,
> When he finds out I published first![7]

Gauss had a long history of investigation of alternative geometries. At age fifteen, he told his friend Heinrich Christian Schumacher that he could develop logically consistent geometries besides the usual Euclidean geometry. Initially, he set out along the road of trying to deduce the parallel postulate from the other four, but eventually reached the same conclusion he had at fifteen, that there were other consistent geometries. In 1824, he wrote to Franz Taurinus, in part to correct an error in Taurinus's purported proof of the parallel postulate. After doing so, Gauss wrote, "The assumption that the sum of the three angles of a triangle is less than 180° leads to a curious geometry, quite different from [the Euclidean], but thoroughly consistent, which I have developed to my satisfaction. . . . The theorems of this geometry appear to be paradoxical and, to the uninitiated, absurd; but calm, steady reflection reveals that they contain nothing impossible."[8] It seems fairly clear that Gauss had not constructed a model for a consistent geometry, but merely had convinced himself that one was possible.

Gauss concluded his letter to Taurinus by saying, "In any case, consider this a private communication, of which no public use or use leading to publicity is to be made. Perhaps I shall myself, if I have at some future time more leisure than in my present circumstances, make public my investigations." Several years later, in a letter to the astronomer Heinrich Olbers, he reiterated both his results and his desire not to go public with them.

Nonetheless, he was sufficiently impressed with the possibility that the geometry of the real world might not be Euclidean that he conceived of an experiment to resolve the matter. Saccheri and Gauss had both deduced that if the parallel postulate did not hold, the sum of the angles in a triangle would total less than 180 degrees. Gauss laid out a triangle using mountains around his home in Göttingen; the sides of the triangle were approximately 40 miles long. He measured the angles of the triangle and computed their sum; had the result been significantly less than 180 degrees, he would have been able to reach an earthshaking conclusion. It was not to be: the sum of the angles differed by less than 2 seconds (1/1,800 of a degree), a difference that could certainly have been the result of experimental error.

Wolfgang and János Bolyai

Wolfgang Bolyai (a.k.a. Farkas Bolyai) was a friend of Gauss from their student days at Gottingen. As students, they had discussed what they referred to as the problem of parallels, and over the years they maintained friendship by correspondence when Wolfgang returned to Hungary. However, Wolfgang's son János was unquestionably the mathematical star of the family. Wolfgang gave János instruction in mathematics, and the son proved to be an extraordinarily quick learner. Wolfgang fell ill one day when János was thirteen, but the father had no qualms about sending in his son to pinch-hit for his lectures at college. I'm not sure how I would have felt if a thirteen-year-old showed up in place of my usual professor.

When János was sixteen, Wofang wrote to Gauss, asking him to take János into Gauss's household as an apprentice to facilitate the advancement of his career. Possibly the letter went astray, but Gauss did not answer, and so János entered the Imperial Engineering Academy, planning on a career in the army. In addition to being an extremely talented mathematician, János was a superb duelist and an enthusiastic violinist. He once accepted a challenge in which he fought thirteen consecutive duels against cavalry officers, but stipulated that he be allowed to play a violin piece after every two duels. He won all thirteen duels; there are no reviews of his violin performances.

While at the academy, János evinced interest in the parallel postulate. Like everyone else, his initial efforts were devoted to trying to prove it. His father, who had battled unsuccessfully with the problem, urged his son to expend his effort elsewhere. "Do not waste one hour's time on that problem," wrote Wolfgang. "It does not lead to any result, instead it will come to poison all your life. . . . I believe that I myself have investigated all conceivable ideas in this connection."[9]

János was not the first son to disregard his father's advice, and in 1823 sent this communiqué to his father: "I am resolved to publish a work on parallels as soon as I can complete and arrange the material, and the opportunity arises. At the moment I still do not clearly see my way through, but the path which I have followed is almost certain to lead me to my goal, provided it is at all possible. . . . All I can say at present is that out of nothing I have created a strange new world."[10]

János had, indeed, created a strange new world. He developed a complete system of geometry, constructing three distinct families of different sets of postulates. The first system incorporated the five classic postulates of Euclid—this is obviously Euclidean geometry. The second system, now known as hyperbolic geometry, included the first four postulates of Euclid and the negation of the parallel postulate. This was to be János's great contribution, a systematic development of non-Euclidean geometry. Finally, his last system, absolute geometry, was based only on Euclid's first four postulates.

János's work, the only thing he ever published, was included as a twenty-four-page appendix to a textbook written by his father. His father sent the work to Gauss, who wrote to a friend that he considered János Bolyai to be a genius of the first order. However, his letter to Wolfgang was quite different. Gauss commented, "To praise it would amount to praising myself. For the entire content of the work . . . coincides almost exactly with my own meditations which have occupied my mind for the past thirty or thirty-five years."[11]

Although this was not intended to be a put-down, it had a devastating effect on János, who was tremendously disturbed that Gauss had earlier traversed the same path. János's life deteriorated significantly thereafter. He received a small pension when he was mustered out of the army and went to live on a family estate. Isolated from the mathematical community, he continued to develop some of his own ideas, and left twenty thousand pages of notes on mathematics behind him. János was to become even more embittered when his towering achievement—being the first person to publish a consistent non-Euclidean geometry—was taken from him.

Nikolai Ivanovich Lobachevsky

The third discoverer of non-Euclidean geometry in this recitation was in actuality the first—or at least the first to publish. Nikolai Ivanovich Lobachevsky was the son of a poor government clerk. His father died when Nikolai was seven, and his widow moved to Kazan in eastern Siberia. Nikolai and his two brothers received public scholarships to secondary schools, and Nikolai entered Kazan University, intending to become a medical administrator. Instead, he would spend the rest of his life there as a student, teacher, and administrator.

He was obviously an extremely talented student, for he graduated from the university before his twentieth birthday with a master's degree in both physics and mathematics. He then received an assistant professorship and became a full professor at age twenty-three. Admittedly, other talented mathematicians have become full professors at an early age, but nonetheless this was an impressive achievement.

Lobachevsky worked along roughly the same lines as Gauss and Bolyai, substituting the assumption that through each point not on a given line there existed more than one parallel to the given line, and going on from there to develop hyperbolic geometry. He published this in 1829 (thus establishing priority, for Gauss never published and Bolyai's effort was published in 1833) in a memoir entitled *On the Foundations of Geometry*. However, instead of publishing it in a reviewed journal it appeared in the *Kazan Messenger*, a monthly house organ published by the university. Lobachevsky, believing that it deserved a wider and more knowledgeable audience, then submitted it to the St. Petersburg Academy—where it was summarily rejected by a buffoon of a referee who failed to appreciate its value. The last sentence may seem rather strong, but having had a few papers bounced in my career by similar buffoons, I can sympathize with Lobachevsky. At any rate, Lobachevsky's effort was another addition to the lengthy list of great papers that initially got bounced.

To Lobachevsky's credit, he refused to be discouraged, and finally had a book published in 1840 in Berlin with the title *Geometric Investigations on the Theory of Parallels*. Lobachevsky sent a copy of the book to Gauss, who was sufficiently impressed to write a congratulatory letter to Lobachevsky. Gauss also wrote to his old friend, Schumacher, with whom he had first discussed alternative geometries, that although he was not surprised at

Lobachevsky's results, having anticipated them, but nonetheless he was intrigued by the methods he had used to derive them. Gauss even studied Russian in his old age so that he could read Lobachevsky's other papers!

Lobachevsky's life differed significantly from János Bolyai's. Lobachevsky became the rector of Kazan University at age thirty-four and lived comfortably thereafter, yet he never ceased his attempts to have his efforts in non-Euclidean geometry recognized. For the fiftieth anniversary of Kazan University, he made one final attempt. Even though he had become blind, he dictated "Pangeometry, or a Summary of the Geometric Foundations of a General and Rigorous Theory of Parallels," which was published in the scientific journal of Kazan University.

Recognition for Lobachevsky would eventually follow, although not until after his death. Just as Hilbert had saluted the efforts of Cantor, the English mathematician William Clifford said of Lobachevsky, "What Vesalius was to Galen, what Copernicus was to Ptolemy, that was Lobachevsky to Euclid."[12] Today, all three of the major participants are recognized as codiscoverers of non-Euclidean geometry, although the bulk of the credit goes to Bolyai and Lobachevsky, who developed their ideas independently—and published them. Sadly, when Bolyai learned of Lobachevsky's work, he initially believed that it was an attempt by Gauss to rob him of his rightful place in the mathematical firmament, and that Gauss had given Lobachevsky some of Bolyai's ideas. Nonetheless, when Bolyai examined Lobachevsky's work, he retained enough integrity to comment that some of Lobachevsky's proofs were the work of a genius, and the entire opus was a monumental achievement.

Another Parallel

It is fascinating how often history repeats itself—even the history of mathematics. We have seen that the story of the continuum hypothesis is much like the story of the parallel postulate. An axiomatic system is outlined, and the status of an additional axiom is in doubt—is it provable from the original axioms, or not? In both cases, the additional axiom turned out to be independent of the original set—the inclusion of either the additional axiom or its negation resulted in consistent systems of axioms. What is just as fascinating is how the stories parallel each other—a great mathematician (Kronecker for Cantor, Gauss for Bolyai) either deliberately (Kronecker) or inadvertently (Gauss) prevents a lesser mathematician from achieving the recognition he deserves, and it is left to posterity to bestow the accolades. Meanwhile, the effect is to ruin a life.

Mathematics is no different from most other human endeavors, in that there are individuals of estimable achievement but substantially less than estimable character.

Eugenio Beltrami and the Last Piece of the Puzzle

There was one final obstacle that had not yet been surmounted: the development of a model that would exhibit the wondrous geometric properties that Gauss, Bolyai, and Lobachevsky had formulated. This was accomplished by Eugenio Beltrami, an Italian geometer, who in 1868 wrote a paper in which he actually constructed such a model. Beltrami was definitely trying to find a concrete realization for the theory that the three early non-Euclidean geometers had developed, for he wrote in this paper that "We have tried to find a real foundation to this doctrine, instead of having to admit for it the necessity of a new order of entities and concepts."[13] Beltrami also played an important role in the history of non-Euclidean geometry, as it was he who first focused attention on the work Saccheri had done.

Many interesting curves in mathematics have resulted from the analysis of a physical problem. One such curve is the tractrix, which is the curve generated by the following situation. Imagine that the string of a yo-yo is completely extended and the free end fastened to a model train traveling on a straight track. The train moves at a constant velocity, keeping the string taut. The tractrix represents the curve traced out by the center of the yo-yo; it gets closer and closer to the track, but never quite reaches it.

If the tractrix is rotated round the railroad track, the track represents a central axis of symmetry for the resulting surface, which is known as a pseudosphere. The pseudosphere is the long-sought model for a non-Euclidean geometry, and every triangle drawn on its surface has the sum of the angles less than 180 degrees.

Is the Universe Euclidean or Non-Euclidean?

Gauss's experiment in measuring the sum of the angles in a triangle whose sides were approximately 40 miles long was the first to try to determine whether the geometry of the universe could be non-Euclidean. Recall that Gauss found, to within experimental error, that his measurement was consistent with a Euclidean universe. This is still a question that fascinates astronomers, and so experiments have continued up through the present day, with the lengths being employed now on the order of billions of light-years. The latest data, from the Wilkinson Microwave Anisotropy Probe, comes down firmly on the side of the Greeks—as best we can determine, the large-scale geometry of the universe is flat, much as it appeared to the Greeks even before it occurred to them that Earth itself might actually be round.

NOTES

1. R. Trudeau, *The Non-Euclidean Revolution* (Boston, Mass: Birkhauser, 1987), p. 30. Items such as points and lines are called primitive terms. Euclid was saying that points are the smallest objects that there are, and cannot be subdivided. He also says such things as "a line is breadthless length," but qualifies it with the adjective "straight" as the situation dictates.
2. Ibid., p. 40. I'm not enough of an expert to be sure that these statements are absolute translations from Greek, but these are basically the ones everyone uses.
3. Ibid., p. 43. You have to wonder why *this* particular version of the parallel postulate was chosen. It seems awfully awkward, and it was no wonder that substitutes were sought. It's usually easier to work with a simple characterization of a concept than a more complicated one.
4. Ibid., p. 128.
5. D. Burton, *The History of Mathematics* (New York: McGraw-Hill, 1993), p. 544.
6. Linus Pauling was once asked how he got so many good ideas. He replied to the effect that he just got a lot of ideas and threw the bad ones away. I attempted to do this, but ran into two obstacles—I didn't have anywhere near the number of ideas that Pauling had, and when I threw away the bad ones, not much was left. There were a few, though.
7. Tom Lehrer received a bachelor's degree in mathematics from Harvard at age eighteen, and a master's degree one year later. Headed for a brilliant mathematical career, he was sidetracked by becoming one of the three greatest humorists of the twentieth century, in my opinion (the other two being Ogden Nash and P. G. Wodehouse). He was probably the first politically incorrect black humorist—yes, before Lenny Bruce and Mort Sahl—and his songs are classics. There will always be a soft spot in my heart for "Nikolai Ivanovich Lobachevsky," "The Old Dope Peddler," and "The Hunting Song," but the one most likely to incite a riot is "I Wanna Go Back to Dixie." Enjoy. Visit http://members.aol.com/quentncree/lehrer/lobachev.htm.
8. Burton, *History of Mathematics*, p. 545.
9. Ibid., p. 548.
10 Ibid., p. 549.
11. Ibid. pp. 549–50.
12. Ibid. p. 554.
13. See http://www-groups.dcs.st-and.ac.uk/~history/Biographies/Beltrami.html.

7
Even Logic Has Limits

Liar, Liar

Back when I was in college and my GPA needed a boost, I would seek out the comforting shelter of the Philosophy Department, which always offered an assortment of introductory courses in logic. The one I took began by examining this classic syllogism:

> All men are mortal.
> Socrates is a man.
> Therefore, Socrates is mortal.

OK, not exactly a deduction requiring a Sherlock Holmes—but there are more interesting formulations that might have intrigued even the great detective. One such syllogism, which did not appear in introductory logic, initially appears to be a clone of the one above:

> All Cretans are liars.
> Epimenides is a Cretan.
> Therefore, Epimenides is a liar.

It seems pretty much the same—unless the first statement was made by Epimenides! If so, is Epimenides lying with the first statement? After all, a liar is one who lies some of the time, but not necessarily all the time. If he is a liar, then the first statement could be a lie—and so some Cretans might not be liars, and we cannot legitimately make the deduction.

There's some wiggle room here; what exactly characterizes a liar? Does he or she need to lie with every statement, or is someone a liar if he occasionally lies? After some refining, the liar's paradox, as this sequence of statements is often called, was condensed to a four-word sentence: this statement is false. Is the statement *This statement is false* true or false? Assuming that true and false are the only alternatives for statements, it cannot be true (if so, it would be true that it is false, and would therefore be false), and it cannot be false (if so, it would be false that it is false, and would therefore be true). Assuming that it is either true or false leads to the conclusion that it is both true and false, and so we must place the sentence *This statement is false* outside the true-false realm. You may be able to detect in this argument the faint echo of the classic odd-even proof that the square root of 2 is irrational, which proceeds by showing that a number simultaneously has two incompatible characteristics.

Some might place the liar's paradox under the heading of "snack food for thought"; on the surface it may seem little more than a curiosity-provoking, but pedantic, point of linguistics. But Kurt Gödel, a talented young mathematician, looked more deeply at the liar's paradox, and used it to prove one of the most thought-provoking mathematical results of the twentieth century.

The Colossus

At the summit of the mathematical world in 1900 perched a colossus—David Hilbert. A student of Ferdinand von Lindemann, the mathematician who had proved that π was transcendental, Hilbert had made brilliant contributions to many of the major fields of mathematics—algebra, geometry, and analysis, the branch of mathematics that evolved from a rigorous examination of some of the theoretical difficulties that accompanied the development of calculus. Hilbert also submitted a paper on the general theory of relativity five days prior to Einstein, although it was not a complete description of the theory.[1] By any standard, though, Hilbert was a titan.

During the International Congress of Mathematicians in Paris in 1900, Hilbert made perhaps the most influential speech ever made at a mathematical meeting. In it, he set the agenda for mathematics in the twentieth century by describing twenty-three critical problems[2]—although, unlike the Clay Institute, he was unable to offer financial incentives for their solution. The first problem on Hilbert's list was the continuum hypothesis, as we have seen this was shown to be undecidable within the Zermelo-Fraenkel formulation of set theory. Second on the list was to discover whether the axioms of arithmetic are consistent.

Recall that an axiomatic scheme is consistent if it is impossible to obtain contradictory results within the system; that is, if it is impossible to prove that the same result can be both true and false. Only one proposition need be both true and false for a scheme to be inconsistent, but it may seem that one could never prove that a proposition that is both true and false does not exist. After all, mustn't one be able to prove all the results stemming from a particular axiom scheme in order to decide whether the scheme is consistent?

Fortunately not. One of the easiest systems of logic to analyze is propositional logic, which is the logic of true/false truth tables. This system, which is frequently taught in Math for Liberal Arts courses, involves constructing and analyzing compound statements that are built up from simple statements (which are only allowed to be true or false) using the terms *not, and, or,* and *if . . . then.* In the following truth table, P and Q are simple statements; the rest of the top line represents the compound statements whose truth values depend upon the truth values of P and Q, and how we compute them. It is rather like an addition table where we use TRUE and FALSE rather than numbers, and compound statements rather than sums.

Row	P	Q	NOT P	P AND Q	P OR Q	IF P THEN Q
(1)	TRUE	TRUE	FALSE	TRUE	TRUE	TRUE
(2)	TRUE	FALSE	FALSE	FALSE	TRUE	FALSE
(3)	FALSE	TRUE	TRUE	FALSE	TRUE	TRUE
(4)	FALSE	FALSE	TRUE	FALSE	FALSE	TRUE

The first two columns list the four possible assignments of the values TRUE and FALSE to the statements P and Q; for example, row 3 gives the truth values of the various statements in the top when P is FALSE and Q is TRUE.

The truth value assigned to NOT P is just the opposite of the truth value assigned to P. As an example, if P is the true statement *The sun rises in the east*, then NOT P is the false statement *The sun does not rise in the east*.

The truth value assigned to P AND Q also reflects common understanding of the word *and,* which requires both P and Q to be true in order for the statement P AND Q to be true. The last two columns require a little more explanation.

The word *or* is used in two different senses in the English language: the exclusive sense and the inclusive sense. When assigning truth values to the statement P OR Q, doing so for the "exclusive or" would result in the statement being true precisely when exactly one of the two statements P and Q were true, whereas doing so for the "inclusive or" would result in the statement being true if at least one of the two statements P and Q were true. The example that I give to Math for Liberal Arts students to distinguish between the two occurs when your waiter or waitress asks you if you would like coffee or dessert after the meal. Your server is going with the inclusive or, because you will never hear a server say, "Sorry, you can only have one or the other," when you say, "I'd like a cup of coffee and a dish of chocolate ice cream." Propositional logic has adopted the "inclusive or," and the table above reflects this.

Finally, the truth values assigned to the statement IF P THEN Q are motivated by the desire to distinguish obviously false arguments: those that start from a true hypothesis and end with a false conclusion. This has a tendency to cause some confusion, because both the following compound statements are defined to be true.

If London is the largest city in England, then the sun rises in the east.

If Yuba City is the largest city in California, then 2+2=4.

The objection students make to the first statement being true is that there's no connecting logical argument, and the objection to the second is that it's impossible to reach the arithmetic conclusion just because the hypothesis is false. IF P THEN Q does not mean (in propositional logic) that there is a logical argument starting from P and ending with Q. One of the original goals of propositional logic was to distinguish obviously fallacious arguments from all others; there's something clearly wrong with an argument that goes *2+2=4, therefore the sun rises in the west.* It's tempting to think of IF P THEN Q as an implication (which means there is some underlying connecting argument), but it's not the way propositional logic regards it.

Propositional logic incorporates a method of computing the true/false value of a compound statement, just as arithmetic can compute a value for $x+yz$ when numerical values of x, y, and z are given. For instance, if P and Q are TRUE and R is FALSE, the compound statement (P AND NOT Q) OR R is evaluated according to the above table in the following fashion.

(TRUE AND NOT TRUE) OR FALSE
(TRUE AND FALSE) OR FALSE
FALSE OR FALSE
FALSE

Finally, just as arithmetic statements such as $x(y+z)=xy+xz$ is universally true because no matter what values of x, y, and z are substituted, both sides evaluate to the same number, it is possible for two compound statements to have identical values no matter what the truth values of the individual statements that make up the compound statements are. In this case, the two statements are called logically equivalent; the truth tables below shows that NOT (P OR Q) is logically equivalent to (NOT P) AND (NOT Q).

Row	P	Q	P OR Q	NOT (P OR Q)
(1)	TRUE	TRUE	TRUE	FALSE
(2)	TRUE	FALSE	TRUE	FALSE
(3)	FALSE	TRUE	TRUE	FALSE
(4)	FALSE	FALSE	FALSE	TRUE

The last column of this truth table has the same values as the last column of the following table.

Row	P	Q	NOT P	NOT Q	(NOT P) AND (NOT Q)
(1)	TRUE	TRUE	FALSE	FALSE	FALSE
(2)	TRUE	FALSE	FALSE	TRUE	FALSE
(3)	FALSE	TRUE	TRUE	FALSE	FALSE
(4)	FALSE	FALSE	TRUE	TRUE	TRUE

A situation in which this equivalence arises occurs when your server asks you if you would like coffee or dessert, and you reply that you don't. The server does not bring you coffee and also does not bring you dessert.

Propositional logic was shown to be consistent in the early 1920s by Emil Post, using a proof that could be followed by any high-school logic student.[3] Post showed that under the assumption that propositional logic was inconsistent, any proposition could be shown to be true, including propositions such as P AND (NOT P), which is always false. The next step was to tackle the problem of the consistency of other systems—which brings us back to the second problem on Hilbert's list, the consistency of arithmetic.

Peano's Axioms

Numerous formulations of the axioms of arithmetic exist, but the ones that mathematicians and logicians use were devised by Giuseppe Peano, an Italian mathematician of the late nineteenth and early twentieth century. His axioms for the natural numbers (another term for the positive integers) were

> Axiom 1: The number 1 is a natural number.
> Axiom 2: If a is a natural number, so is $a+1$.
> Axiom 3: If a and b are natural numbers with $a=b$, then $a+1=b+1$.
> Axiom 4: If a is a natural number, then $a+1 \neq 1$.

If these were the only axioms, not only would you be able to balance your checkbook, but mathematicians would have no difficulty showing that the axioms were consistent. It was Peano's fifth axiom that caused the problems.

Axiom 5: If S is any set that contains 1, and has the property that if a belongs to S, so does $a+1$, then S contains all the natural numbers.

This last axiom, sometimes called the principle of mathematical induction, allows mathematicians to prove results about all natural numbers. Suppose that one day you find yourself at a boring meeting, and with nothing better to do you start jotting down sums of odd numbers. After a short while you have compiled the following table:

$$1=1$$
$$1+3=4$$
$$1+3+5=9$$
$$1+3+5+7=16$$

Suddenly, you notice that all the numbers on the right are squares, and you also notice that the number on the right is the square of the number of odd numbers on the left. This leads you to form the following conjecture: the sum of the first n odd numbers (the last of which is $2n-1$) is n^2. You can write this as a single formula:

$$1+3+5+\ldots+(2n-1)=n^2.$$

So how are you going to prove this? There are at least two cute ways to do this. The first is an algebraic version of Gauss's trick. Write down the sum S in both increasing and decreasing order.

$$S=1 \qquad +3 \qquad +\cdots+(2n-3) \qquad +(2n-1)$$
$$S=(2n-1) \quad +(2n-3) \quad +\cdots+3 \qquad \quad +1$$

Each sum contains precisely n terms, so if we add the left sides of both equations we get $2S$, and by looking at the sums of each column, we notice that $1+(2n-1)=2n=3+(2n-3)$, and so on. Adding the right sides, we get n sums of $2n$, or $2n^2$. So $2S=2n^2$, and the result follows.

The second way is so easy that third graders to whom I've presented talks understand the idea. It requires looking at these sums on a checkerboard. The number 1 is represented by the square in the upper-left-hand corner of the checkerboard. The number 3 is represented by all the squares in the second row or column that share a vertex with the upper-left-hand corner square. Together, $1+3$ makes up the square in the upper left hand corner that is two checkerboard squares on a side. The number 5 is represented by all the squares in the third row or column that share a vertex with a square used in the representation of 3. Together, $1+3+5$ makes up the square in the upper-left-hand corner that is three checkerboard squares on a side. And so on.

You can also use the principle of mathematical induction.

The line

$$1=1^2$$

establishes the proposition (the sum of the first n odd numbers is n^2) for $n=1$. If we assume the proposition is true for the integer n, all we need do is to show that the proposition is true for $n+1$. This proposition would be that the sum of the first $n+1$ odd numbers is $(n+1)^2$. Written formally, we need to establish that, under the assumption

$$1+3+5+\ldots+(2n-1)=n^2 \qquad \text{(the formula is valid for the integer } n\text{)}$$

we can proceed to prove

$$1+3+5+\ldots+(2(n+1)-1)=(n+1)^2 \qquad \text{(the formula is valid for the integer } n+1\text{)}$$

The basic facts of algebraic and arithmetic manipulation can be deduced from the Peano axioms, but to do so is somewhat technical, and so for the remainder of this proof we'll just assume the usual laws of arithmetic and algebra, such as $a+b=b+a$.

Simplifying the expression in parentheses on the left side of the equation yields

$$1+3+5+\ldots+(2n+1)=(n+1)^2$$

Continuing, we obtain

$$1+3+5+\ldots+(2n+1)=[1+3+5+\ldots+(2n-1)]+(2n+1)$$
$$=n^2+(2n+1) \qquad \text{(this substitution is our assumption)}$$
$$=(n+1)^2 \qquad \text{(basic algebra)}$$

If A denotes the set of all positive integers n such that the sum of the first n odd integers is n^2, we have shown that A contains 1, and if n belongs to A, then $n + 1$ belongs to A. By Axiom 5, A contains all the positive integers.

A substantial number of deep results use mathematical induction as a key proof technique. Demonstrating the inconsistency of arithmetic would make a lot of mathematicians very unhappy—including David Hilbert, whose basis theorem (an important result in both ring theory and algebraic geometry) was proved using mathematical induction. It seems fairly safe to say that Hilbert definitely hoped that someone would prove that the Peano axioms for arithmetic were consistent; after all, no one wants to see one of his most famous results put in doubt.

So there is a lot riding on establishing that the Peano axioms for arithmetic are consistent, and Hilbert was well aware of this—that's why it's Problem Number 2, ahead of some truly famous problems like the Goldbach conjecture (every even number is the sum of two primes) and the Riemann hypothesis (a technical result with immense potential, but which requires an acquaintance with complex variables and infinite series to understand it). Suffice it to say that the Clay Mathematics Institute will pay $1 million to anyone who manages to demonstrate the consistency or inconsistency of the Peano axioms.

A Postdoc Shakes Things Up

There is a belief that mathematicians do their best work before they are thirty. Possibly forty would be a more reasonable estimate—the Fields Medal is awarded only for work done prior to that age. Nonetheless, some of the most important results in mathematics have been the work of graduate and postdoctoral students.

There is a good deal of debate on why this should be the case; my own belief is that to some extent, work on a particular problem sometimes becomes ossified, in the sense that the leading mathematicians have blazed a trail that most others follow—and sometimes that trail leads just so far and no further. Young mathematicians are less likely to have been indoctrinated—I recall vividly Bill Bade, my thesis adviser, handing me reading material that would bring me up to date, but not suggesting what line I should pursue after I had finished reading the papers.

Kurt Gödel was born six years after Hilbert propounded his twenty-three problems, in what is now the Czech Republic. His academic talents were apparent from an early age. Gödel initially debated between studying mathematics and theoretical physics, but opted for mathematics because of a course he took from a charismatic instructor who was confined to a wheelchair. Gödel was highly conscious of his own health problems—a consciousness that was later to prove his undoing, so it is possible that the instructor's condition had a significant impact on Gödel's decision.

Mathematicians in Europe generally have to overcome two hurdles on the road to a tenured professorship: the doctoral dissertation (as do American mathematicians), and the habilitation (thankfully not required of American mathematicians), which is an additional noteworthy performance after the doctorate has been awarded. Gödel had become interested in mathematical logic, and his doctoral dissertation consisted of a proof that a system of predicate logic proposed in part by Hilbert was complete—every true result in the system was provable. This result was a considerable leap beyond Post's demonstration that propositional logic was consistent—and Gödel's proof used mathematical induction to establish the result. For his habilitation, Gödel decided to go after really big game—the consistency of arithmetic, number two on Hilbert's list of twenty-three problems.

In August 1930, having completed his work, Gödel submitted a contributed paper for a mathematics conference that featured an address by Hilbert entitled "Logic and the Understanding of Nature." Hilbert was still on the trail of axiomatizing physics and proving arithmetic was consistent, and he ended his speech with supreme confidence: "We must know. We shall know." It is somewhat ironic that Gödel's contributed paper at the same conference contained results, delivered in a twenty-minute talk, that were to dash forever Hilbert's dream of "We shall know." In an address delivered far from the limelight (or what passes for limelight at a mathematics conference), Gödel announced his result that one of two conditions must exist: either arithmetic included propositions that could not be proved (now known as undecidable propositions), or that Peano's axioms were inconsistent. To this day, no one has shown that Peano's axioms are inconsistent, and despite the lingering uncertainty you can get almost infinite odds from any mathematician that they are not. This result is known as Gödel's incompleteness theorem.

Unlike Einstein's theory of relativity, which took the world of physics by storm and was accepted almost immediately, the mathematics community initially did not appreciate the significance of Gödel's work. Nonetheless, during the ensuing five years or so, his results gained widespread recognition and acceptance. He continued to do impressive work in mathematical logic, despite encountering problems in terms of his health. Although Gödel was not a Jew, he could easily have been mistaken for one (he was once attacked by a gang of thugs who thought he was Jewish), and when one of his influential teachers was murdered by a Nazi student in 1936, Gödel suffered a nervous breakdown. When World War II began, Gödel left Germany and traveled to America by way of Russia and Japan, ending up at Princeton.

Health problems, both physical and mental, continued to plague Gödel. His circle of friends and acquaintances at Princeton was very select—there were periods during which the only person to whom he spoke was Einstein. Toward the end of his life, paranoia gained the upper hand, and his health problems led him to believe that people were trying to poison him. He died in 1978 from attempting to avoid being poisoned by refusing to eat.

Proofs of Gödel's Incompleteness Theorem

There are many different ways to go about demonstrating Gödel's theorem. I have elected to go with demonstrating here that it is plausible and have given a reference to formal proof in the notes to this chapter that gives the flavor of Gödel's original proof.[4]

Gödel took the liar's paradox, and modified the sentence *This statement is false* (which, as we have seen, lies outside the realm of statements that can be judged to be true or false) to *This statement is unprovable.* That was Gödel's starting point, but by a technique known as Gödel numbering, which is described in the proof, he linked unprovability of statements to unprovability of statements about integers in the Peano axioms framework. If the statement *This statement is unprovable* is unprovable, then it is true, and the link Gödel established with arithmetic showed that there exist unprovable statements in number theory. If the statement *This statement is unprovable* is provable then it is false, and Gödel's proof linked this conclusion to the inconsistency of the Peano Axioms.[5]

What exactly is meant by the word *unprovable?* It means just what it says, that there is no proof that will determine the truth or falsity of the statement. Needless to say, the existence of unprovable statements raises some questions. There are two schools of thought on the subject. Recall that the uncertainty principle is interpreted by most physicists to mean that conjugate variables do not have specifically defined values, not that humans are just not good enough to measure the specifically defined variables. One group of mathematicians interpret unprovability in the same fashion—it isn't that we aren't bright enough to prove that a statement is true or false, it's that if logic is used as the ultimate arbiter, it is inadequate to the task. Others view an unprovable statement as one that is inherently true or false, but the system of logic used just doesn't reach far enough to discern it.

The Halting Problem

At approximately the same time that Gödel came up with his incompleteness theorem (which perhaps should be called the incompleteness or inconsistency theorem, but this doesn't sound as good), mathematicians were beginning to construct computers, and also to formulate the theory that underlies the process of computation. The first relatively complicated computer programs had been written, and mathematicians discovered a nasty possibility lurking within the computational process: the computer might go into an infinite loop, from which the only rescue was to halt the program manually (this probably meant disconnecting the computer from the power source). Here is an easy example of an infinite loop.

Program Statement Number	Instruction
1	Go to Program Statement 2
2	Go to Program Statement 1

The first instruction in the program sends the program to the second instruction, which sends it back to the first instruction, and so on.

In the early days of computer programming, entering an infinite loop was a common occurrence, and so a natural question arose: Could one write a computer program whose sole purpose was to determine whether a computer program would enter an infinite loop? Actually, the question was phrased differently, but equivalently: If a program either halts or loops, is it possible to write a computer program that determines whether another computer program will halt or loop? This was known as the halting problem.

It was quickly shown that it was not possible to write such a computer program; this result was known as the unsolvability of the halting problem. The following proof is due to Alan Turing, one of the early giants in the field. Turing was not only a tremendously talented mathematician and logician, but he also played a major role in helping decipher German codes during World War II. However, he was a homosexual in an environment tremendously intolerant of homosexuality, and was forced to undergo chemical treatments, which resulted in his eventual suicide.

Suppose that the halting problem is solvable, and there is a program H that, given a program P and an input I, can determine whether P halts or loops. The output from the program H is the result; H halts if it determines that P halts on input I, and H loops if it determines that P loops on input I. We now construct a new program N that examines the output of

H and does the opposite; if H outputs "halt," then N loops, and if H outputs "loop," then N halts.

Since H presumably is able to determine whether a program halts or not, let's take the program N and use it as the input to N. If H determines that N halts, then the output of H is "halt," and so N loops. If H determines that N loops, then the output of H is "loop," and so N halts. In other words, N does the opposite of what H thinks it should do. This contradiction resulted from our assumption that the halting problem was solvable; therefore, the halting problem must be unsolvable.

That probably wasn't so hard to follow. It looks like the proof incorporates elements similar to those that existed in the liar's paradox, and looks are not deceptive in this case. Mathematicians have shown that even though the results appear to be in disparate areas, Gödel's theorem and the unsolvability of the halting problem are equivalent; each can be proved as a consequence of the other.

Flash forward to the present. The unsolvability of the halting problem turns out to be equivalent to a problem whose unsolvability will probably ensure the continued existence of a multibillion dollar industry. The year 2007 marked the twenty-fifth anniversary of the first appearance of a computer virus. The Elk Cloner virus was developed for Apple II computers by Rich Skrenta, a Pittsburgh high-school student, and did nothing more sinister than copy itself to operating systems and floppy disks (remember them?), and display the following less-than-memorable verse on the monitor screen.

> Elk Cloner: The program with a personality
> It will get on all your disks
> It will infiltrate your chips
> Yes it's Cloner!
> It will stick to you like glue
> It will modify RAM too
> Send in the Cloner![6]

Skrenta was to prove no threat to Keats or Frost as a literary figure, but from this humble effort sprung the whole spectrum of malware. It also resulted in the obvious question: Is it possible to write a program that will detect computer viruses? Happily for the continued existence of firms such as Norton and McAfee, computer programs can be written to detect some viruses, but the criminals will always be ahead of the police, at least in this area. The existence of a computer program to detect all viruses is equivalent to the halting problem; no such program can be written.[7]

What Is, or Might Be, Undecidable

I'm not sure what the future holds in this area, but I am sure of what mathematicians would like to see. Just as stock market traders would like to hear a bell ring at a market bottom, mathematicians would like a quick way to determine whether a problem on which they are working is undecidable. Regrettably, Gödel's theorem does not come with an algorithm that tells them precisely which propositions are undecidable. The example Gödel constructed of an undecidable proposition is mathematically useless; it involves formulas whose Gödel number satisfies that formula. The Gödel number of a formula is referenced in the footnotes, but there is not a single mathematically important formula in arithmetic that incorporates the Gödel number of that formula. What mathematicians would really like is a tag attached to such outstanding problems as the Goldbach conjecture, which says either, "Don't bother—this proposition is undecidable," or "Keep at it, you might get somewhere." It seems highly unlikely that anyone will ever find a way to tag all propositions; the history of the field (think of the unsolvability of the halting problem) is that it is far more likely to be shown that no such way to tag propositions exists.

However, some extremely interesting problems have been shown to be undecidable. Unfortunately, there have been relatively few—nowhere near enough to reach some sort of general conclusion as to the type of problem that is undecidable. By far the most important was Cohen's demonstration that the continuum hypothesis was undecidable within Zermelo-Fraenkel set theory (with the axiom of choice) if that theory were consistent. There have been at least two other interesting problems that have been shown to be undecidable—and one is related to a currently unsolved problem that is intriguing and easy to understand.

The Word Problem—No, We're Not Discussing Scrabble

In chapter 5, the group of symmetries of an equilateral triangle was found to consist of combinations of two basic motions: a 120-degree counterclockwise rotation, labeled R, and a move in which the top vertex is unchanged but the bottom two are interchanged, called a flip and labeled F.

If I denotes the identity of the group (the symmetry that leaves all the vertices in their original positions), then we have the following relationships between F and R.

$F^2 = I$ (recall that $F^2 = FF$; two flips of the triangle result in the original position)

$R^3 = I$ (likewise for three successive 120-degree counterclockwise rotations)

$FR^2 = RF$

$R^2F = FR$

As shown in chapter 5, there are a total of six different symmetries of the equilateral triangle, which can be produced by I, R, R^2, F, RF, and FR. Suppose that we used the above four rules to try to reduce lengthy words using only the letters R and F to one of those six. Here's an example.

$$
\begin{aligned}
RFR^2\ FRF \quad &= FR^2\ R^2F\ FR^2 \quad &&\text{(replaced first two and last two)} \\
&= F\ R^3R\ F^2\ R^2 \quad &&(R^2R^2 = R^3R = R^4) \\
&= FIRIR^2 \quad &&(R^3 = F^2 = I) \\
&= FRR^2 \quad &&(FI = F,\ RI = R) \\
&= FR^3 = FI = F \quad &&\text{(Whew!)}
\end{aligned}
$$

It is easy to show that any "word" using just the letters R and F can be reduced by using the three basic relationships to one of the six words corresponding to the symmetries of the equilateral triangle. Here's the game plan: let's show that any string of three letters can be reduced to a string of two or fewer letters. There are eight possibilities; I'll just write out the end result.

$$
\begin{aligned}
RRR &= I \\
RRF &= FR \\
RFR &= F \\
RFF &= R \\
FRR &= RF \\
FRF &= R^2 \\
FFR &= R \\
FFF &= F
\end{aligned}
$$

Since every string of three letters can be reduced to a string of two or fewer letters, keep doing this until you get a word of two or fewer letters; it must be one of the basic six that are the elements of the group. We say that the group S_3 of the symmetries of the equilateral triangle is generated by the two generators R and F subject to the four basic relationships.

There are many (not all) groups that are defined by a collection of generators that are subject to a number of relationships, as in the above example. The word problem for such a group is to find an algorithm that when given two words (such as $RFR^2 FRF$ and RFR), will decide whether they represent the same element of the group. For some groups, this can be done, but in 1955, Novikov gave an example of a group for which the word problem was undecidable.[8] The Novikov family has made great contributions to mathematics. Petr Novikov, of word problem fame, had two sons: Andrei was a distinguished mathematician, and Serge was a very distinguished mathematician, winning a Fields Medal in 1970.

Incidentally, when Rubik's Cube first appeared, lots of papers on solutions for it appeared in journals devoted to group theory[9]—for Rubik's Cube involves a group of symmetries with generators (rotations around the various axes) subject to relationships.

Do You Always Get There from Here?

The last of the three problems that has been shown to be undecidable is known as Goodstein's theorem. To get a feel for this problem, here's a currently unsolved problem, the Collatz conjecture, that has some aspects in common with it. Many mathematicians feel it may be undecidable, but it is easy to understand. Paul Erdos, the prolific and peripatetic mathematician whose lifestyle consisted of visiting various universities for short periods, used to offer monetary prizes for the solutions to interesting problems. Because Erdos was basically supported by the mathematical community, living with mathematicians whose universities he visited, the money he collected for honoraria was used to fund these prizes. They ranged from $10 up; he offered $500 for a proof (one way or the other) of the Collatz conjecture. Of the Collatz conjecture, he said "Mathematics is not yet ready for such problems."[10]

When you first see this problem, it looks like something a nine-year-old kid dreamed up while doodling with numbers. Pick a number. If it is even, divide by 2; if it is odd, multiply by 3 and add 1. Keep doing this. We'll do an example in which 7 is used as the starting number.

$$7,22[=3\times7+1],11[=22/2],34,17,52,26,13,40,20,10,5,16,8,4,2,1$$

It took a while, but we finally hit the number 1. Here's the unsolved problem: Do you always eventually hit the number 1 no matter where you start? If you can prove it, one way or the other, I think that the money is part of Erdos's estate; he died in 1996. The *New York Times* did a front-page article on him after he died. He was once humorously described by his colleague Alfréd Rényi, who said, "A mathematician is a machine for turning coffee into theorems," as Erdos drank prodigious quantities.[11]

Goodstein's theorem[12] is reminiscent of this problem; it defines a sequence (called a Goodstein sequence) recursively (the next term is defined by doing something to the previous term in the sequence, just as in the above unsolved problem), although the sequence it defines is not as simple to state as the one in the Collatz conjecture. It can be shown that every Goodstein sequence terminates at 0—although one cannot show this using the Peano axioms alone; one must use an additional axiom, the axiom of infinity, from Zermelo-Fraenkel set theory. As such, it is an interesting proposition that would be undecidable using only the Peano axioms—as opposed to the uninteresting undecidable propositions used by Gödel in his original proof. It is also worth pointing out that the provability of Goodstein's theorem in a stronger version of set theory lends credence to the point of view that these theorems are inherently true or false, and that an adequate system of logic can determine it.

So, unless another graduate student with the talent of Gödel shows up to prove that the overwhelming consensus of mathematicians is wrong, and that the Peano axioms are actually inconsistent, mathematicians will continue to rely on mathematical induction. It is one of the most useful tools available—and the existence of undecidable propositions is a small price to pay for such a valuable tool. If some graduate student comes along to accomplish such an unlikely task, he or she can be assured of two things: a Fields Medal, and the undying hatred of the mathematical community, whom he or she will have deprived of one of the most valuable weapons in its arsenal.

NOTES

1. See http://www-groups.dcs.st-and.ac.uk/~history/Biographies/Hilbert.html. This may have been the last time in human history when there were polymaths who could make truly significant contributions in several fields. In addition to Hilbert, Henri Poincaré (of Poincaré conjecture fame) also did important work in both mathematics and physics.

2. See http://en.wikipedia.org/wiki/Hilbert's_problems. This contains a list of all twenty-three problems along with their current state. Most of the ones not discussed in this book are fairly technical, but number three is easily understood—given two polyhedra of equal volumes, can you cut the first into a finite number of pieces and reassemble it into the second? That this could not be done was shown by Max Dehn.

3. A. K. Dewdney, *Beyond Reason* (Hoboken, N.J.: John Wiley & Sons, 2004). The proof of the consistency of propositional logic is given on pp. 150–152.

4. Ibid. The proof of the impossibility theorem is given on pp.153–158. See

http://www.miskatonic.org/godel.html. The text box from Rudy Rucker's *Infinity and the Mind* contains Gödel's argument in computer-program form.

5. See http://www.cs.auckland.ac.nz/CDMTCS/chaitin/georgia.html. This site actually has an article that appeared relating Gödel's theorem and information theory. It's pretty close to readable if you're comfortable with mathematical notation.

6. See http://en.wikipedia.org/wiki/Elk_Cloner.

7. *Science* 317 (July 13, 2007): pp. 210–11.

8. See http://www-groups.dcs.st-and.ac.uk/~history/Biographies/Novikov.html.

9. See http://members.tripod.com/~dogschool/. Here's a short course in group theory with good graphics that will get you through the group theory that underlies the Rubik's Cube. The explanation for the reason this site has the whimsical title "The Dog School of Mathematics" can be found by going to the home page.

10. See http://en.wikipedia.org/wiki/Collatz_conjecture. This site has a lot of stuff, much of which can be read with only a high-school background—but not all of it.

11. See http://en.wikipedia.org/wiki/Paul_Erdos. This site gives you a nice picture of Erdos's life as well as his accomplishments.

12. See http://en.wikipedia.org/wiki/Goodstein%27s_theorem. The opening paragraph calls attention to the fact that Goodstein's theorem is a nonartificial example of an undecidable proposition. The mathematics is a little hard to read for the neophyte, but the persistent may be able to handle it.

8

Space and Time: Is That All There Is?

The Second Solution

High-school algebra is more than five decades in my rearview mirror, but the more things change, the more high-school algebra remains pretty much the same. The books are a lot more interesting graphically and a *whole* lot more expensive—but they still contain problems such as the one in the next paragraph.

Susan's garden has the shape of a rectangle. The area of the garden is 50 square yards, and the length of the garden exceeds the width by 5 yards. What are the dimensions of the garden?

The setup for this problem is straightforward. If you let L and W denote the dimensions of the garden, then you have the following equations.

$$LW = 50 \quad \text{(area} = 50 \text{ square yards)}$$
$$L - 5 = W \quad \text{(length exceeds width by 5 yards)}$$

Substituting the second equation into the first results in the quadratic equation $L(L-5) = 50$. Regrouping and factoring, $L^2 - 5L - 50 = 0 = (L-10)(L+5)$. There are two solutions to this equation. One of these is $L = 10$;

substituting this into the second equation gives $W=5$. It's easy to check that these numbers solve the problem; a garden with a length of 10 yards and a width of 5 yards has an area of 50 square yards, and the length exceeds the width by 5 yards.

However, there is a second solution to the above quadratic equation; $L=-5$. Substituting this into the second equation gives $W=-10$, and this pair of numbers gives a satisfactory *mathematical* solution to the pair of equations. Search as you might, though, you're not going to find a garden with a width of negative 10 yards—because width is a quantity that is inherently positive.

The high-school student knows what to do in this case: discard the solution $W=-10$ and $L=-5$, precisely because it is meaningless in the context of the problem. If such an equation were to occur in physics, the physicist would not be quite so quick to cast aside the apparent meaningless solution. Rather, he or she might wonder if there was some hidden underlying meaning to the apparently "meaningless" solution that was yet to be revealed, as there is a rich history of interesting physics underlying apparently meaningless solutions.

The Gap in the Table

The dictionary definition of mathematics is usually similar to the one I found in my ancient Funk & Wagnalls—the study of quantity, form, magnitude, and arrangement. When an arrangement manifests itself so that, in part, it explains phenomena in the real world, exploration is often undertaken to see whether undiscovered phenomena correspond to missing parts of the arrangement. A classic such case is the discovery of the periodic table of the elements.

In the nineteenth century, the chemists were attempting to impose order and structure to the apparently bewildering array of the chemical elements. Dmitry Mendeleyev, a Russian chemist, decided to try to organize the known elements into a pattern. To do so, he first arranged these elements in increasing order of atomic weight, the same physical property that had attracted the attention of John Dalton when he devised the atomic theory. He then imposed another level of order by grouping the elements according to secondary properties such as metallicity and chemical reactivity—the ease with which elements combined with other elements.

The result of Mendeleyev's deliberations was the periodic table of the elements, a tabular arrangement of the elements in both rows and columns. In essence, each column was characterized by a specific chemical property, such as alkali metal or chemically nonreactive gas. The atomic weights increased from left to right in each row, and from top to bottom in each column.

When Mendeleyev began his work, not all the elements were known. As a result, there were occasional gaps in the periodic table—places where Mendeleyev would have expected an element with a particular atomic weight and chemical properties to be, but no such element was known to exist. With supreme confidence, Mendeleyev predicted the future discovery of three such elements, giving their approximate atomic weights and chemical properties even before their existence could be substantiated. His most famous prediction involved an element that Mendeleyev called eka-silicon. Located between silicon and tin in one of his columns, Mendeleyev predicted that it would be a metal with properties resembling those of silicon and tin. Further, he made several quantifiable predictions: its weight would be 5.5 times heavier than water, its oxide would be 4.7 times heavier than water, and so on. When eka-silicon (later called germanium) was discovered some twenty years later, Mendeleyev's predictions were right on the money.

While this may be the most notable success of discovering an arrangement to which the real world conformed in part, and then attempting to discover aspects of the real world that conformed to other parts of the arrangement, this story has been frequently repeated in physics.

The Garden of Negative Width

One of the most famous of these examples occurred when Paul Dirac published an equation in 1928 describing the behavior of an electron moving in an arbitrary electromagnetic field. The solutions to Dirac's equation occurred in pairs, somewhat analogous to the way that the complex roots of a quadratic $ax^2 + bx + c$ whose discriminant $b^2 - 4ac$ is negative occur in complex conjugate pairs, having the form $u + iv$ and $u - iv$. Any solution in which the particle had positive energy had a counterpart in a solution in which the particle had negative energy—an idea almost as puzzling as a garden whose width is negative. Dirac realized that this could correspond to an electron-like particle whose charge was positive (the charge on an electron is negative), an idea initially greeted with considerable skepticism. The great Russian physicist Pyotr Kapitsa attended weekly seminars with Dirac. No matter what the topic of the seminar, at its end Kapitsa would turn to Dirac and say, "Paul, where is the antielectron?"[1]

The last laugh, however, was to be Dirac's. In 1932, the American physicist Carl Anderson discovered the antielectron (which was renamed the positron) in an experiment involving the tracks left by cosmic rays in a cloud chamber. It isn't recorded whether, after the discovery of the positron, Dirac turned to Kapitsa and said, "There!" If Dirac could have resisted the temptation, he would have been one of the rare people able to do so. Dirac shared the Nobel Prize in 1933.

In mathematics, one way to avoid the dilemma posed by the existence of gardens of negative width is to restrict the domain of the function (the set of allowed input values) under consideration. Thus, when considering the equations for the garden described at the outset of this chapter, one might consider only those values of L and W (the length and width of the garden) that are positive. Thus restricted, the quadratic equation we obtained has only one solution in the allowed domain of the function, and the problem of gardens of negative width is eliminated.

However, as in the example of Dirac's equation, the physicist cannot cavalierly restrict the domain of functions that describe phenomena. By doing so, the restricted domain might describe phenomena that are known—but in the part of the domain that was excluded might lurk something unexpected and wonderful.

Complex Cookies

Mathematical concepts are idealizations. Some idealizations, such as "three" or "point," have close correspondences with our intuitive understanding of the world. Some, such as i (the square root of -1) have utility without such close correspondence. A staple mathematical tool in quantum mechanics is the wave function, which is a complex-valued function whose squares are probability density functions. Probability density functions are fairly easy to understand: I am more likely to be in Los Angeles (my home) next Tuesday than I am to be in Cleveland, but there are certainly events that would necessitate a trip there. Low-probability events, to be sure—but not impossible ones. The complex-valued function whose square is a probability density function does not seem to have any correspondence to the world—it is a mathematical entity that, when properly manipulated, gives accurate results about the world.

But what have complex numbers to do with the real world? We cannot buy $2-3i$ cookies for $10+15i$ cents per cookie—but if we could, we could pay the bill! Using the formula that cost equals number of cookies times the price per cookie, the total cost would be

$$(2-3i) \times (10+15i) = 20 + 30i - 30i + 45 = 65 \text{ cents}$$

Analogous situations frequently occur in physics—real phenomena have unreal, but useful, descriptions. Does the utility of these descriptions end in the universe we know—or have we just not discovered complex cookies?

Heisenberg said something about the role of mathematics that seems appropriate to quote at this juncture, even though we aren't talking about quantum mechanics: ". . . it has been possible to invent a mathematical scheme—the quantum theory—which seems entirely adequate for the treatment of atomic processes; for visualisation, however, we must content ourselves with two incomplete analogies—the wave picture and the corpuscular picture."[2] In other words, complex cookies may not be a part of what we can visualize with the accuracy that we can depict them mathematically—but if it works, that's all we need to worry about.

The Standard Model

The Standard Model represents the way physicists currently view the universe. There are two types of particles: the fermions, which are the particles of matter, and the bosons, which are the particles that transmit the four forces currently thought to act in the universe. These forces are electromagnetism, which is transmitted by photons; the weak nuclear force, which is responsible for radioactive decay and is transmitted by W and Z bosons; the strong nuclear force, which holds the nuclei together (counteracting the repulsive electric force generated by the protons in the nucleus) and is transmitted by gluons; and the gravitational force, for which the transmitting particle has yet to be found.

The Standard Model is the culmination of centuries of effort, but even if it is shown to be accurate in every detail (and in some instances it has been experimentally confirmed to more than fifteen decimal places), physicists know that it leaves many questions unanswered. The masses of the particles are numbers that are measured experimentally; is there a deeper theory that can predict those masses? The fermions divide nicely into three separate "generations" of particles; why three, and not two, four, or some other number? The four forces vary greatly in many respects. Electromagnetism is almost forty orders of magnitude stronger than gravity, which is why you can run a comb through your hair (assuming you have some; I don't) on a cold winter day and generate enough static electricity to overcome the gravitational attraction of Earth and pick up a small Post-it. Electromagnetism and gravity have infinite range—the strong force is confined to the interior of the atom. Electromagnetism is both attractive and repulsive, which fortunately is gathered in equal amounts in every un-ionized atom (so we are not walking bundles of electric charge, except on cold winter days), but gravity is always attractive.

Electrical storms in the center of the Milky Way do not affect us, but the gravity that emanates from the black hole at its center most certainly does.

Above all, is there a deeper level to reality than the one shown by the Standard Model? We have already seen that the Aspect experiments have confirmed that there are no hidden variables underlying quantum-mechanical properties, but that merely eliminates a deeper reality in one specific situation.

Beyond the Standard Model

Physics at the moment is awash with myriad variations of Dirac's anti-electron. There are numerous attempts to go beyond the Standard Model (which categorizes the array of particles and forces now thought to comprise our universe) by answering the question, Why these particles and forces? The quest for an elegant theory of everything will undoubtedly continue, because only the discovery of such a theory or a proof that no such theory is possible can derail the quest. As a result, mathematical descriptions that extend the Standard Model are currently abundant. We shall examine some of the consequences of these models—the particles, structures, and dimensions whose existence might possibly never be known.

The Other Side of Infinity

There is probably only one thing on which every mathematical model for physics agrees—in this universe, there is no such thing as infinity.

That's not to say that there is no such thing as infinity in any universe. In an intriguing and provocative article[3] (which initially appeared in *Scientific American*), the physicist Max Tegmark classified four different types of "parallel universes" that could be explored. His Level IV classification consisted of mathematical structures. Tegmark argues cogently, if not necessarily persuasively, for a concept he calls "mathematical democracy." The multiverse (which is the collection of all possible universes) consists of every possible physical realization of a mathematical model.

There is certainly good reason to consider this possibility. The Nobel Prize–winning physicist Leo Szilard, reflecting on the "unreasonable effectiveness" of mathematics in physics, declaring that he could see no rational reason for it. John Archibald Wheeler, whom we encountered in a discussion on the physical utility of the continuum, wondered, "Why these equations?"[4] Unsaid, but implied, was "Why not other equations?" Why does the universe in which we live support Einstein's equations in general relativity and Maxwell's equations in electromagnetism, but not some other set of equations? Tegmark proposes a possible answer: the multiverse supports all possible (consistent) sets of equations; it just does so in different sectors, and we happen to be living in the Einstein-Maxwell sector.

One of the great debates that raged in physics for centuries is the nature of light—is it a wave or is it a particle? The answer to this, that it's both, would not be fully appreciated until the twentieth century, but in the middle of the nineteenth century Maxwell's equations, which described electromagnetic behavior, appeared to give the nod to waves, as the equations led to solutions that were obviously wavelike. Nonetheless, a problem still remained: waves were thought to need a medium in which to propagate. Water waves need water (or some other liquid) and sound waves need air (or some other substance to transmit the alternating rarefactions and compressions that constitute waves). The medium in which electromagnetic waves were believed to propagate was the exquisitely named luminiferous aether. With such a lovely appellation, it was a pity that experiments initially conducted by Albert Michelson and Edward Morley in 1887, and which continue up to the present day, have demonstrated to an extraordinarily high degree of precision that there is no such thing as luminiferous aether. The Michelson-Morley result led quickly to the Lorentz transformations, which expressed the relationships between distance and time in a coordinate frame moving at constant velocity with respect to another frame, and these transformations helped Einstein formulate special relativity, best known for the formula $E=mc^2$. However, Einstein also managed to derive the following expression for mass as a function of its velocity

$$m = \frac{m_0}{\sqrt{1 - (v/c)^2}}$$

Here m_0 is the mass of the object at rest, and m is its mass when it is moving with velocity v.[5] It is easy to see that when v is greater than 0 but less than c, the denominator is less than 1, and so the mass m is greater than the rest mass m_0. Equally easy to see is that as v gets closer to c, the denominator approaches 0, and m gets larger and larger: when v is 90 percent of the speed of light, the mass has more than doubled; when v is 99 percent of the speed, the mass has increased by a factor of 7; and when v is 99.99 percent of the speed of light, the mass is more than 70 times what it was at rest.

As we noted above, our universe (or the physicists currently populating our sector of it) abhors infinities much the same way as it was thought that nature abhorred a vacuum. As a result, no particle with a finite mass can travel at the speed of light—for then the denominator in the above equation would be 0, and the mass m would be infinite. This does not prevent light from traveling at the speed of light, for photons, the particles of light, are massless—they have no rest mass.

Enter the Tachyon?

Einstein's theory also showed that infinite energy would be required to enable a particle with nonzero mass to move at the speed of light. However, a closer look at the above equation—which was derived for objects in our universe—reveals a potential counterpart of Dirac's antielectron. If v is greater than c, the denominator requires us to take the square root of a negative number, resulting in an imaginary number. The rules governing the arithmetic of imaginary numbers dictate that the result of dividing a real number by an imaginary number is an imaginary number, so if it were possible to accelerate an object beyond the speed of light, the mass of the object would become imaginary. Objects with imaginary masses traveling faster than the speed of light are called tachyons—from the Greek word for "speed." No messengers from the other side of infinity have yet been detected in our universe—but absence of evidence is not evidence of absence. Tachyons have such a bad reputation in contemporary physics that a theory that allows them is said to have an instability,[6] but they have not been totally ruled out. While there is no way to envision how a tachyon that "slowed down" to speeds less than that of light could suddenly become a particle with real mass, some particles known to exist do change character; there are three different species of neutrinos, and they change species as they travel. Bizarre as the notion of a particle changing its species appears, it is the only current explanation for what is referred to as the solar neutrino deficit problem. Decades of collecting neutrinos resulted in only one-third the expected number of neutrinos; the only way to account for this is to assume that neutrinos actually change species in flight, as the neutrino collectors could detect only a single species of neutrino.

String Theory

Earlier in this chapter we referred to John Archibald Wheeler's remark, "Why these equations?" An equally valid, and perhaps more down-to-this-universe question, is "Why these particles?" Why are the particles that comprise our universe, the photons and quarks and gluons and electrons and neutrinos of the Standard Model, *the* particles? Why do they have the masses and interaction strengths that they do? These questions are beyond the scope of the Standard Model. The Standard Model is a table of "what" that enables us to predict "how"—but completely unaddressed is the question of "why."

Possibly, "why" is a question lurking beyond the realm of physics—but maybe not. The last century has seen science's view of the fundamental particles change first from atoms to neutrons, protons, and electrons, and then to the particles that comprise the Standard Model. Perhaps there are even more fundamental particles that make up those in the Standard Model. The leading candidate theory in this area is string theory[7] (and a more evolved version, known as superstring theory), which postulates that all the particles in this universe are the vibrational modes of one-dimensional objects known as strings. A violin string of a fixed length and tension can be made to vibrate only in particular patterns. When a violinist draws a bow over a single string, the sound is melodious, rather than the caterwauling of discordant sounds. That is because each vibrational pattern corresponds to a particular note. The strings that comprise string theory can vibrate only in particular patterns—and these patterns are the particles that comprise our universe.

The strings that lie at the heart of these theories are incredibly tiny[8]—direct observation of strings is as difficult as trying to read the pages of a book from a distance of 100 light-years. This certainly appears to rule out the possibility of direct observation, but science does not always need direct observation; often, consequences suffice. Scientists had not actually seen an atom until revealed by the scanning microscopes of the 1980s, but the atomic theory was firmly in place more than a hundred years before that. Much effort is being expended to find predictions that string theory makes that could be experimentally or observationally verified. However, string theory is itself a work in progress, and as it goes through various incarnations (there have been at least four generations of string theory so far), the predictions change.

Nonetheless, string theory generally makes two types of predictions that transcend the Standard Model: it predicts particles that have yet to be observed, and geometrical and topological structures for the universe that remain unverified. Both of these merit a look—not only because they are fascinating in and of themselves, but because it is possible that some future theory may show us that these lead to contradictions, and we shall have to look elsewhere.

I was fortunate to attend a lecture several years ago at Caltech given by

Edward Witten. Witten is a Fields Medalist and a leading exponent of string theory. The lecture was attended by many leading scientists, and there was a question-and-answer period after the lecture. One of the questions directed at Witten was "Do you really believe that this is the way things are?" Witten's answer was unequivocal: "If I didn't believe it, I wouldn't have spent ten years working on it." That convinced me—at the time of the lecture. On my way home, it struck me that centuries earlier Isaac Newton, who had expended ten years working on an explanation for alchemy, might have answered a question regarding the validity of alchemy in the same fashion.

More Posited Particles

There are two major classes of particles that have yet to be detected, but are nonetheless the subject of investigation. The first class of particles consists of those that are part of the Standard Model, but have not yet been detected. The star in this particular firmament is the Higgs particle, which is the vehicle by which mass is imparted to all nonmassless particles (photons, the particles of light, are examples of massless particles). As observed earlier, the Higgs particle seems to remain tantalizingly out of reach of whatever energy range the current generation of particle accelerators can deliver, but many physicists feel it's just a matter of time before one turns up in the snares that have been set for it.

More interesting, from a mathematical standpoint, are the supersymmetric particles. These particles are the ephemeral dance partners for the chorus line of particles making up the Standard Model, and exist in most of the currently popular variations of string theory. Like Dirac's antielectron, they emerge as the result of a pairing process in the underlying mathematics. For Dirac, however, the pairing process resulted from having opposite charge. Supersymmetric particles occur from a pairing involving spin—the mass particles of the Standard Model have spin $1/2$, their supersymmetric particles have spin 0.

The detection of a Higgs particle, or a supersymmetric one, depends upon the mass of these particles. All the unobserved particles are heavy (when measured as a multiple of the mass of the proton); the projected mass of these particles varies with which theory is being employed. What does not vary is what is necessary to create them—lots of energy. Einstein's great mass-energy equation, $E = mc^2$, can be compared to the exchange rate between varying currencies. The atomic bomb, or the energy generated by thermonuclear fusion in the heart of a star, is the result of converting mass into energy—a very little mass generates a lot of energy, because that mass is multiplied by c^2. In order to produce a particle of mass m, one must look at the equivalent equation $m = E/c^2$; and it takes a lot of E to produce a very little m. This means that particle accelerators have to be built ever larger to supply the E necessary to create the new particles; and the larger the m of the new particles, the larger the necessary E. There are string theories in which the masses of the key particles are accessible to the next generation of accelerators—but also there are string theories in which this is not the case. The key parameter is the size of the fundamental entity—the vibrating string—and the smaller the string, the more energy is needed.

The Man of the Millennium

One of my great disappointments of 1999 was *Time* magazine's failure to nominate a Man of the Millennium. It was some consolation that it nominated Einstein as the Man of the Century (good choice!), but it missed a golden opportunity. For me, Isaac Newton was even more of a clear-cut choice for Man of the Millennium than Einstein was for Man of the Century, and there aren't a whole lot of opportunities to nominate a Man of the Millennium.

Isaac Newton is best known for his theory of gravitation, but this is only one of his many accomplishments in both mathematics and physics. However, Newton's biggest accomplishment transcends mathematics and physics, and is the reason that he deserves to be Man of the Millennium: he formulated the scientific method, which helped to kick-start the Industrial Revolution and all that has happened since. The scientific method, as Newton employed it, consisted of gathering data (or examining existing data), devising a theory to explain the data, mathematically deriving predictions from the theory, and checking to see whether those predictions were valid. He did this not only for gravitation, but for mechanics and optics, and transformed Western civilization.

The Man of the Century

Newton's theory of gravitation is unquestionably one of the great intellectual achievements of mankind. It not only explains most everyday stuff, such as the orbits of the planets and the motion of the tides; it is even deep enough to allow for concepts such as black holes, which were at best blue-sky (or black-sky) ideas until a few decades ago. However, the physicists of the late nineteenth century realized that the theory wasn't perfect—some measurements (notably the precession of the orbit of Mercury) differed significantly from the values calculated using Newton's theory.

Einstein did not simply tinker with Newton's theory; he devised a different way of looking at the universe. Nonetheless, both Newton and Einstein visualized a universe in which events could be specified by four numbers (dimensions)—three numbers denoting spatial location, one denoting temporal location. For Newton, however, these four numbers were absolute; all observers would agree on how much spatial distance there was between two events occurring at the same time, and all observers would agree on how much time had elapsed between two events occurring at the same point in space. One of Einstein's contributions was the observation that these numbers were relative; a consequence of Einstein's theory was that moving observers would disagree as to how much time had elapsed between two events occurring at the same point in space. Moving rulers shorten and moving clocks run more slowly, according to Einstein—and an experiment in which two perfectly synchronized atomic clocks were compared, one of which stayed on the ground while the other was flown in a jet around the world, proved that Einstein was right.

Nonetheless, both Newton and Einstein used four numbers to discuss the universe; their universes are four-dimensional. There are two questions that immediately occur. The first—is our universe really four-dimensional?—asks whether Newton, and then Einstein, got it right. The second—are there other, non-four-dimensional universes?—is somewhat deeper, and starts to get into the realm of philosophy (or pure mathematics).

These two questions have occupied physicists for the better part of a century. It is the quest for the theory of everything, by which physicists will not only explain *what* happens, but *why* it happens and *whether* other things can happen (other universes exist) or they can't (ours is the only possible universe). Although a theory of everything will still leave many questions unanswered, such an accomplishment would close the books on one of the great questions of mankind.

The Geometry of the Universe

One of Newton's most famous statements, to be found in his *Principia*, was "I frame no hypotheses; for whatever is not deduced from the phenomena (observational data) is to be called an hypothesis and hypotheses . . . have no place in experimental philosophy. In this philosophy particular propositions are inferred from the data and afterwards rendered general by induction. Thus it was that . . . [my] laws of motion and gravitation were . . . discovered."[9]

He may not have *framed* them, at least for publication, but it is hard to believe that he did not at least *speculate* about them. Among Newton's substantial mathematical achievements was the development of calculus (which was also developed independently by Gottfried Leibniz). The development of Newtonian gravitation in contemporary textbooks is invariably phrased in terms of calculus, as it is so clearly the correct mathematical tool for expressing the results. Interestingly enough, Newton used calculus sparingly in his *Principia*; the great majority of his results were developed using only Euclidean geometry. Newton's ability to use geometry was extraordinary, and it is impossible to believe that once he had expressed the gravitational force between two bodies as varying inversely with the square of the distance between them, he did not speculate upon the connection between this fact and geometry. The fact that the surface area of a sphere is a multiple of the square of the radius was known to the Greek geometers, and if there is a finite amount of "gravitational stuff" emanating from a material body, that gravitational stuff must be spread out over the surface of an expanding sphere. The existence of such gravitational stuff emanating from material bodies in an expanding sphere would explain the inverse square law of gravitation, and surely Newton must have had some thoughts along these lines.

Another Gap in Another Table

The Standard Model is not an equation, but a table. There is a gap in the Standard Model, a particle that fits in perfectly but has not yet been observed. Just as Mendeleyev's organization of the elements led him to predict missing elements and their properties, the gap in the Standard Model cries out to be filled by a boson that transmits the gravitational force (as the other bosons transmit the other forces). This hypothetical particle is known as the graviton.

Gravitons are in some respects a natural way to explain Newton's inverse square law of gravity. Electromagnetism is also a force in which the strength of the attraction or repulsion varies as the inverse square of the distance between electromagnetic particles, and the reason is that photons spread out over the surface of the expanding sphere (expanding at the speed of light) whose center is the source of the emission. If the same number of photons are spread out over two spheres, and the larger has a radius three times the radius of the smaller, the larger sphere has a surface area nine times as large as the surface area of the smaller sphere. Assuming that the same number of photons are used to cover the surface of the sphere, the photon density (which is a measure of the strength of the force) on the surface of the larger sphere is $1/9 = 1/3^2$, the photon density on the surface of the smaller sphere.

Just as Newton almost certainly realized, it is reasonable to postulate a similar mechanism to account for the strength of the gravitational field. But it is here that we encounter one of the major unsolved problems confronting contemporary physics. The three theories that account for the behavior of the nongravitational forces are all quantum theories that account for the forces by describing the behavior of particles. Relativity, the theory that best describes the gravitational force, is a field theory; it speaks of a gravitational field that extends throughout space, and describes the behavior of this field.

This was also the case with Maxwell's equations, the original description of the electromagnetic force. These equations describe how the electric and magnetic fields relate to each other. In the first half of the twentieth century, quantum electrodynamics was invented, which described how the electromagnetic fields were produced as the result of electrically charged particles (which are fermions) interacting by exchanging photons (which are bosons). Quantum electrodynamics served as the model for subsequent quantum theories—the electroweak theory, which provides a unified description of the electromagnetic and weak forces, and the charmingly named quantum chromodynamics, which provides a description of the strong force. However, even though the particle transmitting the gravitational force is in place—at least theoretically—a successful quantum theory of gravity has yet to emerge. The development of this theory is perhaps the most important goal of contemporary theoretical physics.

Lightning in a Bottle

In 1919, Einstein's general theory of relativity was spectacularly confirmed by Eddington's observations of the gravitational deflection of light by the Sun of light from stars. In the same year, Einstein received an extraordinary paper from Theodor Kaluza,[10] a little-known German mathematician.

Kaluza had done something frequently done by mathematicians, but only occasionally by physicists: he had taken well-known results and placed them in a new and hypothetical environment. The well-known results in this case were Einstein's treatment of general relativity; the hypothetical environment in which he placed them was a universe consisting of four space dimensions (rather than the three familiar to us) and one time dimension.

Kaluza probably chose four space dimensions because it is the next step up the complexity ladder from three dimensions. However, Kaluza's approach caught lightning in a bottle—literally. Not only did this assumption result in Einstein's equations of general relativity, which wasn't surprising, but other equations emerged from this treatment—and these equations were none other than Maxwell's equations describing the electromagnetic field.

Every so often, a bizarre assumption results in something totally wonderful and unexpected. Max Planck made a similar bizarre assumption when he postulated that energy came in discrete packets; this assumption resolved many of the existing problems in theoretical physics at the time, even though it was to be years before the assumption was empirically validated. Paul Dirac made a similar assumption about the existence of the antielectron. Kaluza's assumption, and the almost-miraculous simultaneous appearance of the two great theories describing the era's two known forces (gravity and electromagnetism), made a great impression on Einstein. Einstein's enthusiasm was understandable—he spent much of his career in search of a unified field theory that would successfully combine the theories of electromagnetism and gravity. Kaluza's discovery looked like the fast track to such a theory.

There was just one problem: Where was the fourth spatial dimension? Recall Kapitsa asking Dirac, "Paul, where is the antielectron?" The normal three spatial dimensions (north-south, east-west, up-down) seem to suffice to locate any point in the universe. We seem stuck with three dimensions—as it undoubtedly seemed to Kaluza and Einstein. Then a suggestion from the mathematician Oskar Klein appeared to present an attractive possibility for a fourth dimension.

Klein proposed that the fourth dimension was an extremely small one when compared with the usual three dimensions with which we are familiar. The page you are now reading appears to be two-dimensional, but it is in reality three-dimensional; it's just that the thickness (the third dimension) is very small compared with the height and width of the page that comprise the other two dimensions. This suggestion resurrected, at least in theory, Kaluza's four spatial dimensions. However, there still remained the problem that no one had ever seen the fourth spatial dimension, and the state of the art in both theory and experiment were insufficient to the task of exposing it, if indeed it did exist. The Kaluza-Klein theory, as it was called, died a quiet death.

The Standard Model Redux

One of the great discoveries in the last century is the fact that an atom can change its species. Like the shape-shifters of science fiction, species changing allows particles to assume other forms, and species changing among neutrinos accounts for the solar neutrino deficit. But neutrinos are highly standoffish (a neutrino can travel light-years through solid lead without interacting); atoms are the stuff of the real world. An atom that started life as an atom of nitrogen can, through the process known as beta decay, become an atom of carbon. This is one of the many interesting phenomena associated with radioactivity, and is an action promoted by the weak force.

The weak force is weak when compared with the strong force, the force that holds the nucleus of an atom together against the electrical repulsion generated by the protons residing in the nucleus. Although Einstein and Kaluza certainly knew of the phenomena of beta decay, and also were well aware that *something* had to be holding the nucleus together; the weak and strong forces had not been isolated when they were developing their theory.

In the half century between 1940 and 1990, remarkable progress was made in developing the theories of these forces. A theory that combined the electromagnetic force and the weak force was developed by Sheldon Glashow, Abdus Salam, and Steven Weinberg. This theory postulates that at the exceedingly high temperatures that existed in the early universe, these two forces were actually a single force, and the cooling of the universe enabled the two forces to establish themselves as separate forces, in a manner similar to the way different substances in a mixture will precipitate out as the mixture cools. Quantum chromodynamics, the theory of the strong force, was in large part developed by David Politzer, Frank Wilczek, and David Gross. Both these theories, which won Nobel Prizes for their discoverers, have been subjected to experiment and have so far survived; together, they help to comprise the Standard Model of the particles and forces that make up our universe.

The electroweak theory that combines electromagnetism and the weak force is a significant step forward to realizing Einstein's dream of a unified field theory. The current view is that its idea of forces separating as the universe cools is a template for the ultimate unified field theory—for one inconceivably brief moment after the big bang, at some inconceivably high temperature, all the four forces were a single force, and as the universe cooled, they separated out. First to separate out would have been gravity, then the strong force, and finally electromagnetism and the weak force, separated as described by the electroweak theory.

The development of this theory is a work in progress, but it is one that is encountering a major obstacle. The electroweak theory and quantum chromodynamics are quantum theories, which rely heavily on quantum mechanics to produce their extraordinarily accurate results. Relativity, the best theory we have on gravitation, is a classical field theory that makes no mention of quantum mechanics. The experimental levels at which the theories have been confirmed differ remarkably. We can probe subatomic structure at distances of 10^{-18} meter and have found nothing that would contradict the existing electroweak and chromodynamic theories. However, the best we can do to measure the effect of gravity is to confirm it at distances of one-tenth of a millimeter, or 10^{-4} meters. Part of the difficulty is the extraordinary weakness of gravity when compared to the other forces; the gravity of Earth cannot overcome the static electric force when you run a comb through your hair on a cold winter day, and it requires the gravity of a star to tear apart an atom.

Extra Dimensions Resurrected

The advent of string theory resurrected the Kaluza-Klein theory of additional space dimensions—but in a way that seems almost impossible to grasp. After decades of work, string theorists have realized that there is only one possible extra-dimensional space-time that will result in equations compatible with the known universe—but that extra-dimensional space-time requires ten spatial dimensions and one time dimension. If we have as yet been unable to see any evidence of the one extra spatial dimension of the Kaluza-Klein theory, what possible chance do we have of seeing the seven extra ones required by space-time theorists? And what of the extent of these extra dimensions: Are they large, in the sense that the normal three spatial dimensions are large, or are they small—and if so, how small?

Ever since Newton developed the mathematics of calculus to help formulate his theories of mechanics and gravitation, advances in physics have gone hand in hand with advances in mathematics—but there are times that each leads the other. When Maxwell developed his theory of electromagnetism, he used off-the-shelf vector calculus that had been around for nearly a century; and when Einstein came up with general relativity, he discovered that the differential geometry worked out decades previously by Italian mathematicians was just the right tool for the job. However, string theory has been forced to develop much of its own mathematics, and consequently the mathematics—the language in which the results of string theory is expressed—is incompletely understood.

Compounding this problem is one that has affected physics ever since it began relying upon mathematics to phrase its results—the necessity of approximation. When equations cannot be solved exactly—and as we have seen, this is a frequently occurring situation—one possibility is to solve the exact equations approximately, but another is to replace the exact equations with equations that approximate them and solve the approximating equations. Physicists have been doing this for centuries—for small angles, the sine of the angle is approximately equal to its radian measure (just as 360 degrees constitute a full circle, 2π radians do as well), and for most purposes using the angle in an equation rather than its sine results in a much more tractable equation. The equations of string theory sometimes utilize such approximations in order to be solved, and when dealing with something as unknown as infinitesimally small strings and equally infinitesimally small dimensions, it is hard to be certain that the solutions one obtains reflect the way the universe really is.

So how can we tell if string theory, with its ten spatial dimensions, is on the right track? There are two possible approaches, but both are long shots. Confirmation of the existence of strings would constitute inferential proof of the existence of extra dimensions, as mathematical analyses have mandated that the string scenario holds true only in the eleven-dimensional (ten spatial dimensions, one time dimension) universe described above. However, string theory does not unequivocally mandate the *size* of the strings. Although some versions of string theory place the size of the strings in the neighborhood of 10^{-33} meter, which would render them undetectable by any conceivable equipment technology as we know it could muster, there are versions in which strings are (relatively) huge, and possibly detectable, inferentially if not directly, by the next generation of particle accelerators.

The other approach relies upon the fact that the inverse power law that gravity satisfies depends upon the number of spatial dimensions. We see gravity as an inverse square law because, in our three-dimensional universe, gravitons spread out over the boundary of a sphere, whose surface area varies as the square of the radius. In a two-dimensional universe, gravitons would spread out over the boundary of an expanding circle, whose circumference varies directly (is a constant multiple of) the radius. In higher dimensions, the gravitational force would drop precipitously. The boundary of a p-dimensional sphere varies in size as the $(p-1)$st power of the radius, and so we would see an inverse $(p-1)$-power law for gravitation.

That is, if we could measure the gravitational force at distances for which the extra spatial dimensions are significant. The bad news is that the extra spatial dimensions are required by current theory to be no larger than about 10^{-18} meter—and gravity so far can be measured accurately only on scales of about 10^{-4} meter. Thirteen orders of magnitude is a monumental gap, so this is an ultra long shot—more so if the extra spatial dimensions are smaller than 10^{-18} meter.

The Shadow of the Unknowable

While the physics community is pursuing the ultimate theory of reality with both enthusiasm and optimism, it is hard not to reflect on what we have learned during the last century about the limitation of knowledge in the physical universe. There are at least two paths by which the ultimate nature of reality may be something that is forever hidden to us. The first is that the nature of space-time may be so chaotic at Planck length (the length of a string) and Planck time (the time it takes light to travel the length of a string) that we simply cannot measure things accurately enough to determine some critical features of space and time. The second is that the complexity of the axiomatic structure of whatever theory ultimately describes reality admits undecidable propositions—or something similar to them. It may be that those undecidable propositions have no impact on reality—much as the undecidable propositions examined by Gödel were of meta-mathematical, rather than mathematical, interest. On the other hand, it may be that a proposition lurks somewhere that says that the ultimate nature of reality—the "atoms," as it were, of space, time, and matter—are forever beyond our reach. The quest for a theory of everything may well meet the same fate as Hilbert's desire to prove arithmetic consistent. Some mathematician with a solid background in physics, or some physicist who has studied Gödel's incompleteness theorem, may be able to show that a theory of everything cannot exist. Indeed, if someone asked me to bet, this is where I'd place my money.

NOTES

1. See http://physicsweb.org/articles/world/13/3/2. This site is courtesy of *Physics World*, a magazine for physicists and possibly those who are just interested in physics. At any rate, what I've read on the site is quite well written.
2. W. Heisenberg, *Quantum Mechanics* (Chicago: University of Chicago Press, 1930).
3. See http://arxiv.org/PS_cache/astro-ph/pdf/0302/0302131v1.pdf. There is a less-technical version of this (*Scientific American*, May 2003)—but they'll try to sell you a digital subscription. I've been a subscriber for thirty years; it's a great magazine—but this slightly more technical version is *free*. It is also one of the most interesting papers I've read in the past decade. Although portions of it are a little rugged, READ IT, READ IT, READ IT!
4. Ibid.

5. See http://en.wikipedia.org/wiki/Theory_of_relativity. This site is an excellent primer on relativity, and has references that take you much deeper (click on the links to the main articles).

6. B. Greene, *The Fabric of the Cosmos* (New York: Vintage, 2004), p. 502. This is an absolutely wonderful book, as is another book by the same author that will be referenced in note 7. Greene is a top-notch physicist and an expositor with a sense of humor. Nonetheless, there are portions of the book that I had to work to understand. This isn't surprising; this stuff isn't simple. As a local used-car salesman frequently remarked in his TV ads, while pointing out the virtues of a 1985 Chevy, "Flat worth the money!"

7. B. Greene, *The Elegant Universe* (New York: W. W. Norton, 1999). This is the first of the two Greene books—it treats some of the same topics as *Fabric,* but goes much more deeply into relativity and string theory. However, between *Elegant* and *Fabric,* five years elapsed, and a lot happened in string theory, so a reasonable plan is to look at this book first (after all, it was written first), and then follow up with the other.

8. Greene, *Fabric of the Cosmos*, p. 352.

9. I. Newton, *Philosophiae Naturalis Principia Mathematica* (1687). For obvious reasons, everyone refers to it as *Principia*. You know you've encountered a mathematical logician if you say *"Principia"* and he or she thinks you're referring to the classic work in mathematical logic by Betrand Russell and Albert N. Whitehead—best known for the fact that it takes them eight-hundred-plus pages to get around to $1 + 1 = 2$.

10. See http://en.wikipedia.org/wiki/Kaluza. This was basically Kaluza's one moment of glory. Like Cantor, he had a good deal of difficulty getting a professorship in the German university system—despite Einstein's support.

Section III

Information: The Goldilocks Dilemma

Murphy's Law

Nobody knows who Murphy is, but virtually everyone knows Murphy's Law, which sums up much of life's frustrations in seven words: if anything can go wrong, it will.

We've already seen some of the reasons that Murphy's Law has such a trenchant take on reality. On our earlier visit to the garage, we saw that changes that seem to be logical ways to improve an existing situation can actually make things worse. Later, when we look at what chaos theory has to say on the subject, the tiniest, almost-unmeasurable deviation from the plan may cause things to go drastically wrong—for want of a nail, the shoe was lost, and so on.

It's hard for simple things to go wrong. If the only item on your daily agenda is to go to the supermarket and purchase a few basic items, it's awfully hard to foul that up. Yes, the supermarket can be out of something (not your fault), or you may forget something on your list (your fault, but it wasn't the fact that the task was too complicated; your mind was on other matters), but these foul-ups do not arise out of the inherent difficulty of the problem. What mathematics has discovered is that there are some problems that are so intrinsically difficult that it may not be possible to get them right; at least not in a reasonable amount of time.

Another Visit to the Garage

Every so often, we are confronted by an uncomfortably lengthy "to-do" list. Early in life, I adopted the strategy of getting the more onerous chores done first. There were a couple of reasons for this. The first was that at the outset I always had more energy, and the distasteful jobs always require more energy, either physical or emotional. The second was that once the onerous chores are out of the way, I could see the finish line, and this seemed to give me renewed energy for completing the remaining jobs.

I had stumbled on a strategy for scheduling tasks that goes by the name of "decreasing-times processing." If one takes a close look at the schedules that exhibited the unusual anomalies in our earlier trip to the garage, some of the problems were the result of lengthy tasks being scheduled too late. In an attempt to prevent this, the decreasing-times processing algorithm was devised. It consists of constructing the priority list by arranging the tasks in decreasing order of required time (ties are resolved by choosing the task with the smallest number first, so if T3 and T5 require the same time, T3 is scheduled first).

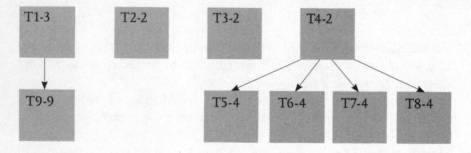

The priority list is T9, T5, T6, T7, T8, T1, T2, T3, T4. With four mechanics, the schedule looks like this.

Mechanic	Task Start and Finish Times					
	0	2	3	6	10	12
Al	T1		T9			Done
Bob	T2	T5		T8	Idle	Done
Chuck	T3	T6		Idle		Done
Don	T4	T7		Idle		Done

There's a lot of idle time here, but that's to be expected. The important point is that all tasks are finished after twelve hours, and that's the optimal solution.

Let's see what happens when we look at the three mechanics when the task times were all reduced by one hour.

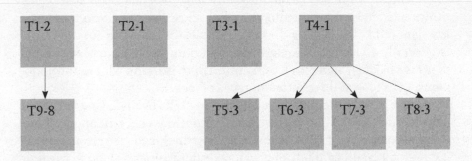

The priority list is the same as the above: T9, T5, T6, T7, T8, T1, T2, T3, T4. This leads to the following schedule.

Mechanic	Task Start and Finish Times					
	0	1	2	5	8	10
Al	T1		T9			Done
Bob	T2	T4	T5	T7	Idle	Done
Chuck	T3	Idle	T6	T8	Idle	Done

Once again, this is the best we can do. Is this the sword that cuts through the Gordian knot of scheduling? Regrettably not. As you might have suspected from the fact that this problem still has a $1 million bounty on its head, neither the priority-list algorithm nor decreasing-times processing will always deliver the optimal schedule. However, decreasing-times processing is superior to the list-processing algorithm in the following important respect: the worst-case scenario with decreasing-times processing is substantially superior to the worst-case scenario with the list-processing algorithm. Suppose that T represents the length of the optimal schedule. If m mechanics are available, then the worst that can happen with the list-processing algorithm is a schedule of length $(2-1/m)T$. However, if decreasing-times processing is used, the worst that can happen is a schedule whose length is $(4T-m)/3$.[1]

There is one such algorithm that always works: construct all possible schedules, and choose the one that best optimizes whatever criteria are used. There's a major problem with that: there could be an awful lot of schedules, especially if there are a lot of tasks.

How Hard Is Hard?

The difficulty of doing something obviously depends upon how many things need to be done. Finding the best schedule to perform four tasks is a slam dunk, but finding the best schedule to perform a hundred tasks is generally a Herculean undertaking. Working with a hundred components is obviously more time consuming than working with four components. Let's look at three different types of jobs.

The first job is something we all do; paying bills by mail. Generally, you have to open the bill, write a check, and put the check in an envelope. Roughly speaking, it takes as much time to do this for the electric bill as it does for a credit card bill, and so it takes four times as long to pay four bills as it does to pay one. Paying bills by mail is linear in the number of components.

The second job is something that happens to all of us: you've put index cards in some sort of order, either in a card box or on a Rolodex, and you drop it. You've got to put the cards back in order. It turns out that this is relatively more time consuming than paying bills, for an obvious reason: as you continue to sort the cards, it takes longer and longer to find the correct place for each additional card.

Finally, there is the scheduling problem. This is even more brutal than sorting the cards for an important reason: all the components must fit together correctly, and you only know whether they fit together correctly when you've finished fitting them together. When you sort the cards, you can hold the last card in your hand and realize that you'll be finished when you've put that card in the correct spot. Such an assessment is not possible with scheduling. As Yogi Berra so famously put it, it ain't over 'til it's over.

Dropping the Rolodex

Suppose we have just dropped our Rolodex on the floor. We now have a bunch of file cards with names, addresses, and phone numbers on them, and we wish to put the cards in alphabetical order. There is a very simple way to do this: take the cards and go through them one at a time, removing each card from the unsorted pile and placing it alphabetically in the new pile by comparing the card we just removed from the old stack with each card in the new stack, one by one, until we find its rightful place. For instance, suppose the top four cards on the old stack, in order, are Betty-Al-Don-Carla. We keep track of the old stack, the new stack, and the number of comparisons that were necessary at each stage.

Step	Old Stack	New Stack	Number of Comparisons for This Step
1	Al-Don-Carla	Betty	0
2	Don-Carla	Al-Betty	1
3	Carla	Al-Betty-Don	2
4		Al-Betty-Carla-Don	3

If there are N cards in the new stack, the maximum number of comparisons that will be needed is N. For instance, at stage 3 above, the card to be compared is Carla, and the old stack is Al-Betty-Don. Carla is behind Al (first comparison), behind Betty (second comparison), in front of Don (third comparison).

We can now look at the worst-case scenario for total number of comparisons. We've seen that the maximum number of comparisons needed is the number of cards in the new stack, and since the new stack builds up one card at a time, with N cards the maximum number of comparisons is $1+2+3+...+(N-1)=N(N-1)/2$, which is a little less than $\frac{1}{2}N^2$. Sorting N cards, even using an inefficient algorithm (better ones are available than the one-at-a-time comparison we used in this example), requires fewer than N^2 comparisons; such a task is said to be doable in polynomial time (we'd say the same thing if we needed fewer than N^4 or N^{12} comparisons). Problems that can be solved in polynomial time as a function of the number of components (cards in the above example) are known as tractable problems. Those problems that can't be solved in polynomial time are called intractable problems.

The Traveling Salesman Problem

This may well be the problem that kicked off the subject of task complexity. Suppose that a salesman has a bunch of different cities to visit, but he starts and ends at his home base. There is a table giving the distance between each city (or, in today's more hectic world, the travel times or possibly the costs); the goal is to devise a route that starts at home and ends there, visits all cities once, and minimizes total travel distance (or travel times or costs).

Let's start by looking at the number of different possible routes we could take. Suppose there are three cities other than home, which we will label generically as A, B, and C. There are six different available routes.

$$\text{Home} \rightarrow A \rightarrow B \rightarrow C \rightarrow \text{Home}$$
$$\text{Home} \rightarrow A \rightarrow C \rightarrow B \rightarrow \text{Home}$$
$$\text{Home} \rightarrow B \rightarrow A \rightarrow C \rightarrow \text{Home}$$
$$\text{Home} \rightarrow B \rightarrow C \rightarrow A \rightarrow \text{Home}$$
$$\text{Home} \rightarrow C \rightarrow A \rightarrow B \rightarrow \text{Home}$$
$$\text{Home} \rightarrow C \rightarrow B \rightarrow A \rightarrow \text{Home}$$

Mercifully, there is a fairly easy way to see how many different routes there are in terms of the number of cities. There are six ways to order the three cities, as we listed above; think of $6 = 3 \times 2 \times 1$. If we have to arrange four cities, we could place any of the four cities first, and arrange the other three in $3 \times 2 \times 1$ ways. This gives a total of $4 \times 3 \times 2 \times 1$ ways of arranging four cities; mathematicians use the factorial notation 4! to abbreviate $4 \times 3 \times 2 \times 1$. The number of ways of arranging N cities in order is $N!$; the argument is basically the one used to show that four cities can be arranged in 4! different orders. So if the traveling salesman must visit N cities, the number of different routes he could take is $N!$

As N gets larger, $N!$ eventually dwarfs any positive power of N, such as N^4 or N^{10}. For instance, let's compare a few values of $N!$ with N^4.

N	N^4	N!
3	81	6
10	10,000	3,628,800
20	160,000	2.43×10^{18}

No matter what power of N we choose to compare with $N!$, $N!$ always swamps it, although when we compare $N!$ with higher powers, such as N^{10}, it takes higher values of N before this phenomenon starts showing up.

Greed Is Not Always Good

Nothing makes us happier than having the easy solution to a problem being the best solution; but, unfortunately, the world is so constructed that this is rarely the case. There is an "easiest" way to construct a passable algorithm for the traveling salesman problem; this algorithm is known as the nearest neighbor algorithm. Whenever the salesman is in a particular town, simply go to the nearest unvisited town (in case of ties, go in alphabetical order). If there are N towns (other than home), it's easy to see that we have to find the smallest of N numbers to find the nearest neighbor, then the smallest of $N-1$ numbers to find the nearest unvisited neighbor to the first town, then the smallest of $N-2$ numbers to find the nearest unvisited number to the second town, etc. So the worst that could happen is that we had to examine a total of $N+(N-1)+(N-2)+\ldots+1=N(N+1)/2$ numbers. Like the card-by-card comparison algorithm when we dropped the Rolodex, this is an algorithm such that the time taken is on the order of N^2, where N is the number of towns.

The nearest neighbor algorithm is an example of what is known as a "greedy" algorithm. There's a technical definition of "greedy algorithm,"[2] but it's pretty clear here what's going on; it's an attempt to construct a path doing the least possible work while still doing some work (simply grabbing the first number available would do the least possible work). Greedy algorithms sometimes give reasonable solutions, but often greed, like crime, doesn't pay.

On our first trip to the garage, we discovered that there are situations in which upgrading all the equipment actually results in a longer completion time. There's an analogous situation for the traveling salesman problem using the nearest neighbor algorithm; it is possible to shorten all the intracity distances and end up with a longer overall trip.

Let's look at a distance table with three towns other than the hometown.

	Home	A	B	C
Home	0	100	105	200
A	100	0	120	300
B	105	120	0	150
C	200	300	150	0

This is a mileage chart similar to the ones found on road maps that used to be available at gas stations. The nearest town to home is A, the nearest unvisited town to A is B, then the salesman must go to C and then home. The total distance for this trip is $100+120+150+200=570$. Suppose we have a slightly different mileage chart in which all intracity distances have been reduced.

	Home	A	B	C
Home	0	95	90	180
A	95	0	115	275
B	90	115	0	140
C	180	275	140	0

Now the nearest town to home is B, the nearest unvisited town to B is A, then the salesman goes to C and returns home. The total distance is $90 + 115 + 275 + 180 = 660$. Of course, this example was constructed to illustrate how greed can be one's downfall; the proximity of B to home lured us into taking a path that is significantly shorter than the best path. If we go first to A, then to B, then to C, and return home, the total distance is $95 + 115 + 140 + 180 = 530$, a considerable improvement. This also shows that the nearest neighbor algorithm doesn't always give us the shortest overall trip.

The traveling salesman problem is considerably simpler than the scheduling problem, in the sense that all one needs are the intracity distances—none of this stuff about digraphs and whether tasks are ready and lists. There is a fairly obvious way to improve on the nearest neighbor algorithm using a technique called "look-ahead." Instead of greedily grabbing the shortest distance, we can use a little foresight and look for the route that minimizes the total distance traveled to the next two towns we visit, rather than just the distance to the next town.

We have to pay a price for this improvement. If there are N towns, we can visit the first two towns in $N \times (N-1)$ ways, then the next two towns in $(N-2) \times (N-3)$, the next two towns in $(N-4) \times (N-5)$ towns, and so on. The total number of distances we must examine is therefore $N \times (N-1) + (N-2) \times (N-3) + (N-4) \times (N-5) + \ldots$. Each of the products in these expressions contains a monomial term N^2, and there are approximately $N/2$ such products, so the total number of distances we must examine is on the order of $N^3/2$. A similar argument shows that if we "look ahead" by computing the shortest distance for the next k towns we visit, the computational overhead is on the order of N^{k+1}.

When You've Solved One, You've Solved Them All

Part of the appeal of mathematics is that the solution of one problem frequently turns out to be the solution of others. Calculus is replete with such situations—one such example is that finding the slope of the tangent to a curve turns out to solve the problem of determining the instantaneous velocity of a moving object when we know its position as a function of time.

In this chapter, we've taken a close look at three problems: schedule construction, the traveling salesman problem, and the card-sorting problem. We've shown the last of these is tractable, but we have yet to determine the status of the other two—although, as Han Solo said in *Star Wars* just before the walls of the garbage-crunching machine started closing in, we've got a really bad feeling about them. They look as if they are intractable, and this would be bad news, as it would mean that we would be confronted with problems of considerable practical significance that simply can't be solved in a reasonable period of time.

Schedule construction and the traveling salesman problem are only a few of more than a thousand such problems that are currently in limbo with regard to whether they are intractable. However, as the result of the work of Stephen Cook, there is a surprising unifying theme that connects all these problems. If you solve one of them, in the sense of finding a polynomial-time algorithm, you've solved them all.

In the 1960s, the University of California, Berkeley was an exciting place to be. Mario Savio was leading the Free Speech Movement in front of Sproul Hall. I was putting the finishing touches on my thesis (although it must be admitted that historians of this era usually neglect to mention this seminal event). Finally, two assistant professors in mathematics and computer science were to become famous: Theodore Kaczynski (later to be known as the Unabomber) and Stephen Cook.

What Stephen Cook did was to connect a wide variety of problems (including the schedule construction problem and the traveling salesman problem) by means of a transformational technique. He discovered an algorithm that, when applied to one of these problems, would change the problem into the form of the other in polynomial time. So, if you could solve the traveling salesman problem in polynomial time, you could transform the schedule construction problem in polynomial time into a traveling salesman problem, which you could also solve in polynomial time.[3] Two successive polynomial-time algorithms (one for transforming the first problem into the second, one for solving the second) comprise a polynomial-time algorithm: for example, if one has a polynomial, such as $P(x) = 2x^2 + 3x - 5$, and we substitute another polynomial, such as $x^3 + 3$, for x in that expression, the result—$2(x^3 + 3)^2 + 3(x^3 + 3) - 5$—is still a polynomial, though admittedly of higher degree.

This greatly raises the stakes for determining whether (or not) there is a polynomial-time algorithm for schedule construction. If you can find one, using Cook's transformational methods, you will have polynomial-time algorithms for more than a thousand other useful problems. You will not only gain undying fame, you will also make lots of money by handling all these problems for a fee—plus, you get to collect one of the Clay Mathematics Institute Millennium Prizes of $1 million for finding such an algorithm. If you can demonstrate that no such algorithm exists, you still get the fame and the $1 million. It's a mystery why people are still trying to trisect the angle when it's known to be impossible, when they could be trying to find a polynomial-time algorithm for the traveling salesman problem and become rich and famous for doing so.

Cook's Tough Cookies

Cook came up with his idea in the early 1970s; by the late 1970s, more than a thousand problems were known to be every bit as difficult to solve as the scheduling problem or the traveling salesman problem. Admittedly, many of these are minor variations of one problem, but it is worth looking at some of the problems to realize how pervasive these really tough problems are.

Satisfiability. This is the problem that Cook first examined. Recall that propositional logic worked with compound statements such as IF (P AND Q) THEN ((NOT Q) OR R). There are three independent variables in this proposition: P, Q, and R. The problem is to determine whether there is an assignment of the values TRUE and FALSE to the variables P, Q, and R such that the compound statement above is TRUE. It's not too hard to see that all you need to do is let P be FALSE, then P AND Q must be false, and any implication in which the hypothesis is FALSE must be TRUE.

The problem is that lengthier compound statements cannot be eyeballed so easily.

The knapsack problem. Imagine that we have a collection of boxes with different weights, and inside each box is an item with a given value. If the knapsack can contain only a maximum weight W, what is the maximum value of the contents of boxes that can be placed inside the knapsack? There are two attractive greedy algorithms here. The first is to sort the items in terms of decreasing value and start stuffing them into the knapsack, most valuable item first, until you can't stuff any more inside. The second is to sort the items in terms of increasing weight and start stuffing them into the knapsack, lightest first, until you are forced to stop.

Remember moneyball, the idea that a team could be put together by maximizing some quantity, such as most home runs hit last year per dollar of current salary? There is a version of that which applies to the knapsack problem. One might sort the items in terms of decreasing value per pound; this strategy might be described as "rare postage stamps first," as I believe rare postage stamps are the most valuable item on the planet as measured in dollars per pound.

Graph coloring. The diagram used to illustrate tasks at the garage is called a digraph, which is short for directed graph. A digraph is a collection of vertices (the task squares in our diagram) with arrows connecting some of them to indicate which tasks must be performed before other tasks. Instead of drawing arrows, which indicate a direction, we might just draw lines connecting some of the vertices. This is very much like an intracity road map, with cities represented by hollow circles at the vertices and lines (which are called edges) connecting the cities indicating major highways (or not so major ones, if you're out in a rural area). A graph is a collection of vertices and edges; two vertices may or may not be connected by an edge, but two cities cannot be connected by more than one edge.

Suppose we decided to fill in each of the hollow circles with a color, subject only to the following rule: if two vertices (the hollow circles) are connected by an edge, they must be colored differently. Obviously, one way to do this is simply to color each city a different color. The graph coloring problem is to determine the minimum number of colors needed to color vertices connected by an edge with different colors.

Mathematicians always like to point out how the most seemingly abstract problem can have unexpected practical applications. The graph coloring problem has lots of these. One such rather surprising application is the assignment of frequencies to users of the electromagnetic spectrum, such as mobile radios or cell phones. Two users who are close to each other cannot share the same frequency, whereas distant users can. The frequencies correspond to the colors.

The Big Question

The big question in this area, one of the Clay Mathematics Institute's million-dollar babies, is whether the tough problems that have been described in this section can be done in polynomial time. Interestingly enough, while an affirmative answer to this question would mean that speedy ways of scheduling or planning routes for the traveling salesman exist (at least in theory; we'd still have to find them), a negative answer would have an upside as well! There is one very important problem for which a negative answer would be highly satisfactory: the factorization problem.

The problem of whether an integer can be factored is, like scheduling and graph coloring, known to be one of Cook's tough cookies. If no polynomial time algorithm exists for doing so, those of us who have bank accounts can breathe a little easier, because as described in the introduction, the difficulty of factoring numbers that are the product of two primes is key to the security of many password-protected systems.

The Experts Weigh In

In 2002, William Gasarch took a poll of a hundred leading experts in this area, asking the question whether the class P of problems solvable in polynomial time was equal to the class NP of Cook's tough cookies. The envelope, please.[4]

> Sixty-one voted that $P \neq NP$ (no polynomial-time algorithm exists for any tough cookie).
>
> Nine voted for $P = NP$.
>
> Four thought that it was an undecidable question in ZFC!
>
> Three thought that it could be resolved by demonstrating an explicit way to solve one of the tough cookies in polynomial time, rather than merely showing an algorithm must exist.
>
> Twenty-two respondents wouldn't even hazard a guess.

Gasarch also asked the respondents to estimate when the problem would be solved. The median guess was in 2050, almost forty-eight years after the poll was taken.

Here are a couple of views from the two opposing camps.

Bela Bollobas: 2020, $P = NP$. "I think that in this respect I am on the loony fringe of the mathematical community. I think (not too strongly) that $P = NP$ and this will be proved within twenty years. Some years ago, Charles Read and I worked on it quite a bit, and we even had a celebratory dinner in a good restaurant before we found an absolutely fatal mistake. I would not be astonished if very clever geometric and combinatorial techniques gave the result, without discovering revolutionary new tools."

Richard Karp: $P \neq NP$. "My intuitive belief is that P is unequal to NP, but the only supporting arguments I can offer are the failure of all efforts to place specific NP-complete problems in P by constructing polynomial-time algorithms. I believe that the traditional proof techniques will not suffice. Something entirely novel will be required. My hunch is that the problem will be solved by a young researcher who is not encumbered by too much conventional wisdom about how to attack the problem."

Notice that one person feels that standard methods suffice, while the other feels that it will require someone who can "think outside the box." I'd vote for the latter; my impression of the history of difficult problems is that many more of them seem to succumb to new approaches rather than to pushing current ideas as far as they will go.

DNA Computers and Quantum Computers

A poll of computer scientists showed that the vast majority believe that no polynomial-time algorithm will be found—but the majority of experts has been wrong many times in the past. Even if they are right, there are some viable alternatives that are being explored.

All the algorithms that we have investigated in this section have been implemented sequentially—for instance, when exploring all the possible routes in the traveling salesman problem, we envision a computer that examines the $N!$ routes one at a time. Another way of handling the problem is to break up the problem into smaller, more manageable chunks, and hand each chunk to a different computer. This is known as parallel computing, and it holds the possibility of greatly speeding up computation. There are several ways to accomplish this outside the realm of using standard computers.

The first of these is DNA computing, which was first achieved by Leonard Adleman of the University of Southern California in 1994 (it's not just a football factory). The idea is to use the ability of strands of DNA to select from a multitude of possible strands the one strand that complements it. Since a quart of liquid contains in the neighborhood of 10^{24} molecules, there is the possibility of greatly speeding up computation—although it's not a viable possibility for *very* large problems.

A potentially more powerful technique is quantum computing, which uses the uniquely quantum phenomenon of superposition to perform massively parallel operations. In a classical computer, which uses 1s and 0s, a 3-bit register always records a definite 3-digit binary integer, such as 110 ($=4+2=6$ as a decimal integer). However, a 3-qubit (a qubit is a quantum bit) register exists in a superposition of all eight 3-digit binary integers, from 000 (decimal integer 0) to 111 (decimal integer 7). Consequently, an N-qubit register exists in a superposition of 2^N states; under the right circumstances, the wave collapse can realize any one of these 2^N possibilities. Since qubits can be very small indeed (possibly even subatomic), a 100-qubit register can encompass 2^{100} different states (approximately 10^{30}), and 100 subatomic particles do not take up a whole lot of space.

While the possibilities for quantum computers are extremely exciting, there are major problems to overcome. One of these is the decoherence problem—the environment tends to react with quantum computers and initiate wave collapse. One would want wave collapse to reveal the answer, rather than having wave collapse occur as the result of random environmental interactions, and so the computer must be kept isolated from the environment for significantly longer periods of time than have presently been achieved.

Settling for Good Enough

Both DNA computing and quantum computing reach into the physical universe for assistance in solving a mathematical problem. This is the reverse of the way matters usually proceed—normally, mathematics is used to solve a problem in the physical universe. Barring a bolt from the blue in the form of a Clay Millennium Prize, the most useful approach is to develop approximate solutions—as we have seen, this is an important area in applied mathematics. For instance, there are algorithms that can find solutions to the traveling salesman problem that are within 2 percent of the best solution, and do so in a reasonable period of time. However, approximate solutions for one problem are not readily transformable into approximate solutions for another problem—for example, the decreasing-times algorithm for the scheduling problem is generally only within 30 percent of the best solution. The fact that the problems that Cook showed to be equivalent appear to require separate approximate solutions is part of the charm—and frustration—of mathematical research. Maybe the next great result in this area is an algorithm for transforming approximation techniques for one of Cook's tough cookies into approximation techniques for the others such that the transformed approximation is within the same percentage of the best solution as the original approximation.

NOTES

1. COMAP, *For All Practical Purposes* (New York: W. H. Freeman & Co., 1988). As I've already remarked, I think this is a terrific book; ideal for people who like mathematics, and pretty good for people who can't stand it but have to take a course in it. Estimation is an extremely important part of mathematics. These are examples of worst-case estimates. Worst-case estimating is also valuable because it often highlights precisely which situations result in the worst case, which can lead to better algorithms.
2. http://mathworld.wolfram.com/search/?query=greedy+algorithm&x=0&y=0.
3. A. K. Dewdney, *Beyond Reason* (Hoboken, N.J.: John Wiley & Sons, 2004). Dewdney shows how to transform the satisfiability problem into the vertex cover problem (a problem in graph theory) by showing how to transform logical expressions into graphs. I don't think this is a general template for transformational techniques. My impression is that there are a bunch of hubs, serving the same function for this area of mathematics that hub airports do for air travel; to show that problem A is transformable into problem B, one transforms problem A to a hub problem, and then the hub problem into problem B.
4. http://www.math.ohio-state.edu/~friedman/pdf/P-P10290512pt.pdf.

10
The Disorganized Universe

The Value of the Unpredictable

It might be thought that true unpredictability would be an absolute barrier to knowledge. While the unpredictability of the random and near random is a source of uncertainty with respect to individual events, the analysis of aggregations of random events is the domain of the subjects of probability and statistics, two of the most highly practical mathematical disciplines. We can only acquire information about the long-term averages of the next flip of the penny; yet information of that type suffices to provide a major foundation stone for our civilization.

Although we never give the matter much thought, on any day there is always the chance that we may suffer or cause an injury while in a car. The absence of insurance would probably not deter any of us from driving, although we would risk financial devastation if either of those two events occurred and we were unable to pay for them. Insurance enables us to avoid such devastation, because we can pay a reasonably small premium to protect ourselves against such an outcome. Insurance companies compile detailed records in order to determine what to charge a middle-aged male driver with one accident in the past five years who wants to insure his 2005 Honda Civic. I sometimes have to bite my lip when contemplating my auto insurance bill, especially in view of the fact that there is a teenage driver in the family. Balancing this is the realization of how different my life would be (if I even had a life to live) had not merchants gathered in coffeehouses in the seventeenth century in order to share jointly the cost of voyages of exploration and commerce. In a sense, aided by the increased accuracy in risk assessment resulting from developments in probability and statistics, we are still doing that today.

Random Is As Random Does

Part of the reason for the success of mathematics is that a mathematician generally knows what other mathematicians are talking about, which is not something you can say about just any field. If you ask mathematicians to define a term such as *group*, you are going to get virtually identical definitions from all of them, but if you ask psychologists to define *love*, you will probably get several variations that depend upon the school of psychological thought to which the respondent adheres.

The shared vocabulary of mathematics is not necessarily esoteric. Most people have just as good a take on some ideas as do mathematicians. If you were to ask John Q. Public for a definition of the word *random*, he would probably say something like "unpredictable." Somewhat surprisingly, the mathematical definition of the term *random variable* goes outside the realm of mathematics into the real world for its definition; a "random variable" is a mathematical function that assigns numbers to the outcomes of random experiments, which is a procedure (such as rolling a die or flipping a coin) in which the outcome of the procedure cannot be determined in advance. Mathematicians use the term *nondeterministic*, which sounds a lot more erudite than *unpredictable*—but both words basically boil down to the same thing. *Deterministic* means that future events depend on present and past ones in a predictable way. *Nondeterministic* events are ones that cannot be so predicted.

But is rolling a die or flipping a coin truly random, in the sense that it is absolutely unpredictable? If one rolls a die, the initial force on the die is known, the topography over which the die is traveling is known, and the laws of physics are the only ones in play, might it not be possible, in theory anyway, to predict the outcome? Obviously, this is a tremendously complicated problem, but the potential gain for gamblers in the world's casinos makes this a tempting problem to solve. In the middle of the twentieth century, a gambler spent years developing a method of throwing the dice in which they spun frantically but did not tumble; this method was so profitable that the gambler was banned from casinos, and it is now a rule in craps games that both dice must hit the wall, which contains numerous bumps that presumably randomize the outcome of the throw.

But does it? If we roll a fair die, will the number 1 (and all the other numbers) come up one-sixth of the time? After all, it seems reasonable that once the die is thrown, only one possible outcome is in accordance with the laws of physics and the initial conditions of the problem—how the gambler held the die, whether his hands were dry or damp, and so on. And so, if the universe knows what's going to happen, why shouldn't we?

Let's grant this argument, temporarily, that given sufficient information and sufficient computational capability, we can determine the outcome of a thrown die. Does that leave anything that can be said to be perfectly random—in the universe, or in mathematics?

One possibility that occurs to us is the randomness that appears in quantum mechanics, but randomness in quantum mechanics, though it has been confirmed to an impressive number of decimal places, is still an infinite number of decimal places short of perfectly random. Maybe mathematics can deliver something ultimately and perfectly random, something that we cannot, under any circumstances, predict.

The Search for the Ideal Random Penny

Let's try to construct a sequence of flips for a penny that conforms to our intuitive idea of how a random penny should behave. We would certainly expect that an ideal random penny should occasionally come up heads three times in a row—and also occasionally (but much more rarely) come up heads 3 million times in a row. This leads us to the realization that there must be an infinite sequence of flips in order to determine whether or not the penny is truly random. Notwithstanding that there are certain technical problems in what we mean by "half" when dealing with an infinite set (those familiar with probability can think of it as having a probability of 0.5), we can try to construct such a sequence. If we use H to denote heads and T to denote tails, the sequence H,T,H,T,H,T,H,T . . . obviously satisfies the restriction that half the flips are heads and half are tails. Equally obviously, it is not a random sequence; we know that if we keep flipping, sooner or later heads or tails will occur twice in a row, and they never do in this sequence. Not only that, but this sequence is perfectly predictable, which is about as far from perfectly random as one can get.

OK, let's modify this sequence, so that each of the possible two-flip pairs (heads-heads, heads-tails, tails-heads, tails-tails) occurs one-quarter of the time. The following sequence will do that.

H,H,H,T,T,H,T,T, H,H,H,T,T,H,T,T, H,H,H,T,T,H,T,T, . . .

In case it isn't clear what's going on here, we repeat the pattern H,H,H,T,T,H,T,T (heads-heads, heads-tails, tails-heads, tails-tails) endlessly. This satisfies two requirements: heads and tails each occur half the time, and each of the two-flip possibilities occurs one-quarter of the time. And yet we still don't have a random sequence; many easily conceivable patterns have been left out. Tails, for example, never occurs three times in a row, as it certainly would given infinitely many flips, and almost certainly would, even after only a hundred.

There is a surprisingly deep question contained here: Can you construct a sequence of flips that is perfectly in accord with the laws of probability, in that each specific sequence of N flips will occur $\frac{1}{2}^N$ of the time?

Number Systems: The Dictionaries of Quantity

The decimal number system (also known as the base-10 number system) that we learn in elementary school is similar to a dictionary. Instead of the letters of the alphabet, the decimal number system uses the characters 0,1,2,3,4,5,6,7,8,9. From these ten characters, it forms all the words that can be used to describe quantity. It's an amazingly simple dictionary; for example, the number 384.07 is actually defined as the sum $3 \times 10^2 + 8 \times 10^1 + 4 \times 10^0 + 0 \times 10^{-1} + 7 \times 10^{-2}$, where the expression $10^{-2} = \frac{1}{10^2}$. The quantitative value of the word 384.07 is deducible from the "letters" used and their positions in the word. I tell my prospective elementary school teachers that it's a lot simpler dictionary than the Merriam-Webster, where one can't deduce the meaning of the word from the letters that make it up, and you have to decide in a split second whether *duck* means "quacking waterfowl" or "watch out for rapidly approaching object."

One way to define the real numbers is the set of all decimal representation of the above form, where we are only allowed to use finitely many numbers to the left of the decimal point but infinitely many thereafter. With this convention, $384.07 = 384.0700000. \ldots$ The rationals are all those numbers, such as $25.512121212 \ldots$, that eventually settle down into a repetitive pattern to the right of the decimal point. A calculator will show (or you can do it by hand) that $.5121212 \ldots = \frac{507}{990}$.

Instead of the "10" that we use in the decimal system, it is possible to use any positive integer greater than 1. When "2" takes the place of "10" in the decimal system, the result is the binary, or base-2, number system.

The alphabet for the binary number system consists of the digits 0 and 1, and a number such as 1011.01 is the binary representation for the number $1 \times 2^3 + 0 \times 2^2 + 1 \times 2^1 + 1 \times 2^0 + 0 \times 2^{-1} + 1 \times 2^{-2}$. Writing each of these terms in the familiar decimal system, this number is $8 + 0 + 2 + 1 + 0 + \frac{1}{4}$ $= 11.25$. The binary system is the natural one to use in storing information in a computer. Originally, information was displayed by means of a sequence of lights: when the light was on, the corresponding digit was a 1; off corresponded to 0. Thus, a row of four lights in the order on-off-off-on corresponded to the binary number $1001 = 1 \times 2^3 + 0 \times 2^2 + 0 \times 2^1 + 1 \times 2^0$, whose decimal value is $8 + 0 + 0 + 1 = 9$. Computers now store information magnetically: if a spot is magnetized, the corresponding digit is a 1; if it is not magnetized, the corresponding digit is a 0.

There is a simple correspondence between an infinite sequence of coin flips and binary representations of numbers between 0 and 1. Given a sequence of flips, simply replace H with 0 and T with 1, remove the commas, and stick a decimal point to the left of the first digit. The infinite sequence of coin flips that alternates heads and tails (H,T,H,T,H, . . .) becomes the binary number .01010. . . . One can also execute this procedure in reverse, going from a binary number between 0 and 1 to an infinite sequence of coin flips. The search for the ideal random penny thus morphs into a search for a binary number. The requirement that each specific sequence of N flips will occur $\frac{1}{2}^N$ of the time becomes the requirement that each specific sequence of N binary digits (0s and 1s) will occur $\frac{1}{2}^N$ of the time. A number that possesses this property is said to be normal in base 2. We'll take a closer look at such numbers shortly.

The Message in Pi

The number pi plays an important role in Carl Sagan's best-selling novel *Contact*,[1] about man's first encounter with an advanced civilization. One of the chapters, entitled "The Message in Pi," outlined the idea that the aliens had delivered a message for mankind hidden deep in the gazillion digits of the expansion of pi. Possibly the fact that pi is known to be a transcendental number prompted Sagan to posit a transcendental message hidden in pi. The lure of mysticism in pi appeared more recently in the movie *Pi* and has doubtless appeared other places of which I am unaware.

It seems awfully unlikely that an alien civilization has so manipulated the geometry of the universe in which they evolved that the ratio of the circumference of a circle to its diameter contains a message. In fact, it seems impossible, considering that the truths, and the parameters, of plane geometry are independent of wherever it is studied. Not only that, the question arises as to how one would translate the message in pi. The digits used in expressing pi are a function of the base of the number, and there would have to be a dictionary that translates blocks of digits into characters in the language in which the message is presented. For example, ASCII is the code that translates blocks of eight binary digits stored in a computer into printable characters; the number 01000001 (whose decimal representation is 65) corresponds to the character *A*. In one sense, however, Sagan was right: mathematicians believe that pi contains not only whatever message the aliens encoded, but every message, repeated infinitely often!

A normal number in base 10 is one in which, on average, each decimal digit, such as 4, appears $\frac{1}{10}$ of the time—but each pair of successive decimal digits, such as 47, appears $\frac{1}{100}$ of the time (there are 100 such pairs, from 00 to 99), each triple of successive decimal digits, such as 471, appears $\frac{1}{1,000}$ of the time, and so on. This is the mathematical equivalent of the "ideal random penny" for which we searched a little while ago, except that instead of an ideal random penny with two sides whose tosses would generate a normal number in base 2, we would imagine a perfectly balanced roulette wheel with 10 numbers, 0 through 9. It is possible to formulate an equivalent definition of normality for any number base. For example, a number that is normal in base 4 is one such that each of the 4^N possible N-digit sequences occurs $\frac{1}{4}^N$ of the time.

Are there any normal numbers, in any base, that we have actually found? Obviously, we cannot flip pennies (perfect or not) forever. There are a few that are known. David Champernowne, who was a classmate of Alan Turing (whose proof of the unsolvability of the halting problem appears in chapter 7), constructed one in 1935. This number, known as Champernowne's constant, is normal in base 10. The number is .123456789101112131415 . . . , which is formed simply by stringing together the decimal representations of the integers in ascending order. When I saw this result, I jumped to the conclusion that Champernowne's constant[2] was also normal in other bases—after all, it's an idea, rather than a specific number, and I naively assumed that whatever proof method worked to show that it was normal in base 10 would work for other bases as well. If there's a *Guinness Book of Records for Most Conclusions Erroneously Jumped To,* I'm probably entitled to an honorable mention. If you look at Champernowne's constant in base 10, it's between .1 and .2—but in base 2, it's a different number. In binary notation, $1=1$, $2=10$, $3=11$, $4=100$, $5=101$, $6=110$, $7=111$, so Champernowne's constant in base 2 starts off .11011100101110111. . . . Any number whose base 2 representation starts off .11 . . . is bigger than $\frac{3}{4}$ (just as .12 . . . in base

10 shows that the number is bigger than $1/10 + 2/10^2$, .11 . . . in base 2 shows that the number is bigger than $1/2^1 + 1/2^2$).

However, in 2001, the base-2 incarnation of Champernowne's constant was shown to be normal in base 2. Thus, writing H for every appearance of 0 and T for every appearance of 1 would be an example of a sequence of tosses that appear to come from a perfectly random penny.

There are a very few known examples of numbers that are normal in every base; all the ones that are known are highly artificial.[3] By "highly artificial," I mean that you are not going to encounter the number in the real world. Obviously, we encounter numbers such as 3.089 (the current price in dollars of a gallon of gasoline in California) and the square root of 2 (when finding the length of the diagonal of a square one foot on a side), but numbers such as Champernowne's constant simply don't show up when we measure things. Numbers that are normal in every base do not appear in the real world, but the real line is chock-full of such numbers. Borel's normal number theorem[4] states that if you pick a real number at random (there's that word again), you are almost certain (in a sense that can be made mathematically specific) to pick a number that is normal in every base. In ordinary usage, when a person is asked to pick a number, he or she will usually pick a number that measures something, such as 5. When a mathematician describes picking a random real number, he or she envisions a process somewhat like a lottery, in which all the real numbers are put into a hat, the hat is thoroughly shaken, and a number is picked out by someone wearing a blindfold. If one does that, the number that is picked out will almost certainly be normal in every base. Again, "almost certainly" has a highly technical definition, but one can get an idea of what is meant by realizing that if a real number is picked at random in the sense described above, it is almost certain that it will not be an integer. Integers form what is known as a set of Lebesgue measure zero; the technical statement of Borel's normal number theorem is that all numbers except for a set of Lebesgue measure zero are normal in every base.

Transcendentals such as pi seem to be prime candidates for numbers that are normal in every base. If pi were shown to be such a number, then Sagan would have been right: the message from the aliens would be encoded in the digits of pi, as the encoded message would simply be a sequence of digits. However, every message is a sequence of digits, so if you dig deep enough into pi, you will find the recipe for the ultimate killer cheesecake, as well as your life story (even the part that hasn't happened yet), repeated infinitely often.

Sagan used to talk about how we are all made of star stuff; the explosions of supernovas create the heavier elements that are used to construct our bodies. He would undoubtedly have been just as intrigued by the intricate way we are all connected to the real line. If a real number is selected at random, it is almost certain that the digits of that number tell the complete stories of every human being who has ever lived, or will ever live, and each story is told infinitely often.

The ideal random penny, whose flips can be viewed as the binary digits of a number that is normal in every base, turns out to be not just a tool for deciding who should kick off and who should receive in the Super Bowl, but an oracle of more-than-Delphic stature. It answers every question that could ever be answered, if only we knew how to read the tea leaves. But of course, we never will.

Tumbling Dice: Why We Can't Know What the Universe Knows

Earlier in the chapter, we asked whether the roll of a die was unpredictable; after all, if the universe knows what is going to happen, why can't we? In the latter portion of the twentieth century, a new branch of mathematics emerged. Chaos theory, as it was to be called, emerged as it was discovered that unpredictable phenomena come in two flavors: inherently unpredictable phenomena, and phenomena that are unpredictable because we cannot obtain sufficient information. Inherently unpredictable phenomena exist only in an idealized sense—the flips of an ideal random penny correspond to the binary digits of a number that is normal in every base, but such a number does not correspond to any quantity that can actually be measured.

The phenomenon of chaos, as it appears in both mathematics and physics, is a specific type of deterministic behavior. Unlike random phenomena, which are completely unpredictable, chaotic phenomena are in theory predictable. The mathematical laws underlying the phenomena are deterministic; the relevant equations have solutions, and the present and past completely determine the future. The problem is not that the laws themselves result in unpredictable phenomena, it is that *we* cannot predict the phenomena because of information underload. Unlike quantum mechanics, where we cannot know the value of parameters because those values do not exist, we cannot (as yet) know the value of parameters because it is impossible for us to gather the requisite information.

Now You're Baking with Chaos

The dictionary defines *chaos* as "a state of extreme confusion and disorder." The recent profusion of restaurant-based reality shows and movies portray the kitchen in such a fashion—harried chefs screaming at waitresses and ingredients destined for the entrée ending up in the dessert. So, giving a nod to popular culture, we turn to the chaotic bedlam of the kitchen for a look at some sophisticated chaos mathematics you can do with a rolling pin.

Imagine that one has a cylinder of dough, and we roll it out so that it is twice as long as it was originally, and then cut it in half and place the right half on top of the left half. We then repeat this rolling and cutting process, which is called the baker's transformation. Think of the pie dough as occupying the segment of the real line between the integers 0 (where the left end of the dough is located) and 1 (where the right end is located). The baker's transformation is a function $B(x)$, which tells us where a point that was originally located at x is located after the rolling, cutting, and placing. $B(x)$ is defined by

$$B(x) = 2x \qquad 0 \leq x \leq \frac{1}{2}$$
$$B(x) = 2x - 1 \qquad \frac{1}{2} < x \leq 1$$

A simple description is that a point on the left half of the dough moves to twice its distance from the left end after the baker's transformation, but a point on the right half doubles its distance from the left end after the rolling, and then is moved one unit toward the left end after the cutting and placing.

This doesn't look very complicated, but surprising things happen. Two points that are originally located very close together can end up very far apart quite quickly. I've chosen two different points that are initially very close to each other, and to honor my wife, who was born on September 1, 1971, her birthday is the first starting point, $x = .090171$. The second starting point is located at .090702, only $\frac{1}{1,000,000}$ of a unit to the right of the first starting point.

After one iteration, the two points have drifted .000002 apart, and even after twelve iterations they are only about .004 apart. But after the sixteenth iteration, one of the points is in the left half of the dough, and the other is now in the right half. The next iteration moves them widely apart—the first point is not far away from the right end, whereas the second point is very near the left end.

Start	Iteration						
	1	12	13	14	15	16	17
0.090171	0.180342	0.340416	0.680832	0.361664	0.723328	0.446656	0.893312
0.090172	0.180344	0.344512	0.689024	0.378048	0.756096	0.512192	0.024384

This example shows that two points that start out very close together can, after a limited number of iterations, occupy positions that are quite distant from one another. This phenomenon has serious implications with regard to using mathematics to predict how systems will behave.

If we were to multiply all the numbers in the above chart by 100, so we could think of them as representing temperature, we can interpret the chart as follows: we have a process in which if we start with a temperature of 9.0171 degrees, after seventeen iterations the temperature is 89.3312 degrees; whereas if we start with a temperature of 9.0172 degrees, after seventeen iterations we end up with a temperature of 2.4384 degrees. Unless we are in a laboratory exercising exquisite control over an experiment, there is no way we can measure the temperature accurately to .0001 degree. Thus, our inability to measure to exquisite accuracy makes it impossible to render accurate predictions; small initial differences may result in substantial subsequent ones. This phenomenon, one of the centerpieces of the science of chaos, is technically known as "extreme sensitivity to initial conditions," but the colloquial expression "the butterfly effect" describes it far more picturesequely: whether or not a butterfly flaps its wings in Brazil could determine whether there is a tornado in Texas two weeks later.

A careful examination of the baker's transformation reveals that it is the cutting process that introduces this difficulty. If two points are both on the left half of the dough, the baker's transformation simply doubles the distance between them—similarly for two points on the right half of the dough. However, if two points are very close but one is on the left half of the dough and the other on the right, the point on the left half ends up very near the right end, but the point on the right half ends up very near the left end. The baker's transformation is an example of what is called a discontinuous function—a function in which small differences in the variable can result in large differences of the corresponding function values. Although discontinuous functions occur in the real world—when you turn on a light, it instantaneously goes from zero brightness to maximum brightness—it may be argued, with some justification, that natural physical processes are more gradual. When the temperature cools, it does not drop from 70 degrees to 50 degrees instantaneously, like the lightbulb—it goes from 70 degrees to 69.9999 degrees to 69.9998 degrees . . . to 50.0001 degrees to 50 degrees.[5] This is a continuous process; small increases in time result in small changes in temperature. Nothing chaotic there, right?

Chaos in the Laboratory

The butterfly effect was actually discovered in conjunction with continuous processes. The development of the transistor made reasonably priced computers available in the late 1950s and early 1960s. Formerly, computers had been hugely expensive arrays of power-hungry vacuum tubes, but by the early 1960s, all universities and many businesses had purchased computers. Business, of course, was using the computers to speed up the computations and store the data needed for commerce, but the universities were using computers to explore computationally intensive problems that were previously inaccessible.

Dr. Edward Lorenz, a professor at MIT, began his career as a mathematician, but later turned his attention to the problem of describing and forecasting the weather. The variables involved are governed by differential equations and systems of differential equations,[6] which describe how the rates at which variables change are related to their current values. These equations, though quite complicated, are associated with continuous processes.

Solving differential equations is an important part of science and engineering, because these are the equations that reflect the behavior of physical processes. However, rarely can one obtain exact solutions to differential equations. As a result, the industry standard approach is to use numerical methods that generate approximate solutions, and numerical methods are most effectively implemented by computers.

One day in 1961, Lorenz programmed a system of differential equations into a computer that probably computed at less that one-tenth of 1 percent of the speed of whatever happens to be sitting on your desk at the moment. As a result, when the time came for lunch, Lorenz recorded the output, turned off the computer, and grabbed a bite. When he returned, he decided to backtrack a little, and did not use the most recent output of the computer, but the output it had generated some iterations previously. He expected the output from the second run to duplicate the output of the previous run (after all, they were running the same iterations), but was surprised to see that after a while, the two sets of outputs differed substantially.

Suspecting that there was either a bug in the program (this happened frequently) or a hardware malfunction (this happened more frequently in 1961 than it does today), he checked both possibilities assiduously—only to find that neither was the case. Then he realized that in reinitializing the computer for the second run, he had rounded off the computer output to the nearest tenth; if the computer said that the temperature was 62.3217 degrees, he had rounded it off to 62.3 degrees. In those days, one had to type in all data by hand, and rounding things off would save substantial typing time. Lorenz figured, quite naturally, that rounding off should make little difference to the computations—but as we saw in the table on page 177, even a difference in the sixth decimal place can cause significant changes in the values of later iterates, at least in the baker's transformation. Lorenz was the first to document and describe a butterfly effect in a system in which variables changed gradually rather than discontinuously. Lorenz is also responsible for the term *butterfly effect*. At a 1972 meeting of the American Association for the Advancement of Science, he presented a paper titled "Predictability: Does the Flap of a Butterfly's Wings in Brazil Set Off a Tornado in Texas?" Later investigations were to reveal that chaotic behavior frequently arose from nonlinear phenomena, a common feature of many important systems. Linear phenomena are those in which a simple multiple of an input results in a like multiple of the output. (Hooke's law is an example of a linear phenomenon. Apply 2 pounds of force to a spring and it stretches 1 inch; apply 8 pounds of force to the spring and it stretches 4 inches.)

Extreme sensitivity to initial conditions was to be a much more pervasive phenomenon than originally suspected. Once chaotic behavior had been described, it was not so surprising that complicated systems such as the weather were subject to the butterfly effect. In the mid-1980s, though, it was shown that the orbit of the now-demoted planet Pluto was also chaotic.[7] The clockwork universe of Newton, in which the heavenly bodies moved serenely in majestic and predictable orbits around the sun, had given way to a much more helter-skelter scenario. Pluto turns out to be eerily similar to the electron in Heisenberg's uncertainty principle; we may know where it is, but we don't know where it's going to be. Well, not really: we don't know where Pluto is going to be because we don't know where it and the other bodies in the solar system are (and how fast and in what direction they are moving) with sufficient accuracy.

Strange Developments

Many systems exhibit periods of stability separated by episodes of transition between these periods. The geysers at Yellowstone Park are a good example. Some, like Old Faithful, are very regular in the timing of their eruptions; others are more erratic. A well-studied example in mathematical ecology is the interplay between the relative populations of predator and prey, such as foxes and rabbits. The dynamics of how the fox and rabbit populations change is qualitatively straightforward. In the presence of an adequate supply of food for the rabbits, the rabbit population will expand, providing more prey for the foxes, whose population will also expand. The foxes will prey on the rabbits, reducing the rabbit population. This reduction will reduce the survival rate of the foxes, enabling the rabbit population to expand again—and so on.

The logistic equation models the relative populations of predator and prey. It has the form $f(x) = a \, x \, (1-x)$, where a is a constant between 0 and 4, and x is a number between 0 and 1 that represents the rabbit fraction of the total population (rabbits divided by the sum of rabbits and foxes) at a given time. The value of the constant a reflects how aggressive the predators are. Imagine that we contrast two different types of predators: boa constrictors and foxes. Boa constrictors have slow metabolisms; a few meals a year keep them satisfied. Foxes, however, are mammals, and need to eat much more frequently in order to survive.

Suppose that x is the rabbit fraction of the population at a given time; then $f(x)$ represents the rabbit fraction of the population one generation later. This new value of $f(x)$ is used as the rabbit fraction of the population to compute the new rabbit fraction after the next generation. Suppose, for example, that $f(x) = 3 \times (1-x)$, and that at some moment $x = .8$ (80 percent of the population consists of rabbits, 20 percent of foxes). Then $f(.8) = 3 \times .8 \times .2 = .48$, so one generation later the rabbits constitute 48 percent of the population. We then compute $f(.48) = 3 \times .48 \times .52 = .7488$, so two generations later the rabbits constitute 74.88 percent of the population.

A fraction x is called an equilibrium point if the rabbit fraction of the population either stays at x or periodically returns to x. It's not too difficult to see why the value of a might change the equilibrium points. If the only predators around are boa constrictors, the rabbit fraction of the total population would undoubtedly be much higher than if the predators were foxes, who burn food quickly and need to eat a lot more often than boa constrictors. Back in the 1980s, when computer monitors had amber screens and blinking white rectangular cursors, there used to be a software simulator for the logistic equation: a program called FOXRAB.[8] While others played Pong on computers, I used to spend time watching FOXRAB, which simply output numbers representing the fraction of the total population that consisted of rabbits.

One might expect the system to evolve smoothly as the constant a gradually increases from 0 to 4, a small increase in a resulting in a small change in the equilibrium points, but the number of equilibrium points of the system behaves very unusually. If a is less than 3, the system has only one equilibrium point; the relative populations eventually remain the same over time. For instance, if $a = 2$, the equilibrium point is $x = .5$; if the population ever consists of 50 percent rabbits, then $f(.5) = 2 \times .5 \times .5 = .5$, and after the next generation (and every subsequent generation) there will be 50 percent rabbits. Other values of x drift toward .5 as time passes. For instance, if $x=.8$, then $f(.8)=2\times.8\times.2=.32$, $f(.32)=2\times.32\times.68=.4352$, and $f(.4352)=.49160192$; after just three generations, a rabbit population of 80 percent has become a rabbit population of almost 50 percent.

At $a=3$, there are two equilibrium points. This state of affairs continues until $a=3.5$, when there are four points; but as a increases to 3.56, the number of equilibrium points increases to eight, then sixteen, then thirty-two, ... When $a=3.569946$, something utterly bizarre happens: there are no equilibrium points at all! As a increases from 3.6 to 4, we see the development of chaos; the number of equilibrium points vary unpredictably, with intervals characterized by an absence of equilibrium points followed by intervals in which the smallest change in the value of a creates a wildly different number of equilibrium points. It's a completely deterministic system, but it's one in which it's impossible to predict the number of equilibrium points. In a chaotic system such as this one, the equilibrium points are called strange attractors.

The following table gives an indication of how the number of equilibrium points of the system changes as a increases. The numbers on the top line represent the generations; the values in the table indicate the fraction of the population consisting of rabbits. In each case, the first generation starts off with half of the total population consisting of rabbits; the rest of the table shows the rabbit fraction for generations 126–134. When $a=2.8$, the population stabilizes at 64.3 percent rabbits. When $a=3.1$, the rabbit population oscillates between 76.5 percent and 55.8 percent. When $a=3.5$, there are four equilibrium points; and when $a=3.55$, there are eight equilibrium points (generation 135 repeats the value of generation 127, generation 136 repeats the value of generation 128, and so on).

Generation	1	126	127	128	129	130	131	132	133	134
$a=2.8$	0.5	0.643	0.643	0.643	0.643	0.643	0.643	0.643	0.643	0.643
$a=3.1$	0.5	0.765	0.558	0.765	0.558	0.765	0.558	0.765	0.558	0.765
$a=3.5$	0.5	0.383	0.827	0.501	0.875	0.383	0.827	0.501	0.875	0.383
$a=3.55$	0.5	0.355	0.813	0.54	0.882	0.37	0.828	0.506	0.887	0.355

The Prevalence of Chaos

Chaotic behavior can be seen in a wide variety of phenomena: the relative populations of predator and prey, the spread pattern of disease epidemics, the onset of cardiac arrhythmia, prices in energy markets, the flipping of the climate between periods of benign temperature and ice ages. Climate flipping is one of the reasons that many scientists are concerned about the phenomenon of greenhouse warming. The climate record contains periods where there have been relatively abrupt transformations, and it is not at all clear what causes the climate to seesaw from one temperature regime to another. Those who feel that humans must take measures to prevent global warming point to the fact that it is impossible to know whether the relative fraction of carbon dioxide in the atmosphere is the trigger of chaotic behavior, but until we are more knowledgeable, it seems prudent to err on the side of caution. On the other side of the argument, the climate seems to have had its own strange attractors for millions of years before man began using fossil fuels as a power source, so we're just Johnny-come-latelies in cycles that have been going on for millions of years without us.

There's a lot to gain by being able to model chaotic systems. Imagine how valuable it would be to be able to predict cardiac arrhythmia before it actually shows up. We've actually taken an important step by knowing that cardiac arrhythmia is a chaotic phenomenon rather than a random one. If it were random, there would be no hope of doing anything about individual cases; the best we could do is to know what percentage of people displaying certain patterns would be liable to suffer heart attacks. With chaotic behavior, there is the possibility that we can do things in individual situations. This probably lies some distance in the future, though, as chaos is a very young discipline.[9] But at least it's not a discipline characterized by extreme confusion and disorder.

NOTES

1. C. Sagan, *Contact* (New York: Simon & Schuster, 1985).
2. See http://mathworld.wolfram.com/NormalNumber.html. Like many of the references in Mathworld, you have to be a pro to take full advantage of the information, but the basics are reasonably comprehensible.
3. See http://mathworld.wolfram.com/AbsolutelyNormal.html.
4. Borel's normal number theorem states that the set of numbers that are not normal in every base is a set of Lebesgue measure zero. You need an upper-division math course to be really comfortable with Lebesgue measure, but it attaches numbers to sets that generalize the idea of length. The Lebesgue measure of the unit interval, all real numbers between 0 and 1, is 1, as you would expect. However, the Lebesgue measure of all the rational numbers in that interval is 0. The proof of Borel's normal number theorem uses the axiom of choice. Probability for sets of real numbers is closely tied up with Lebesgue measure, so when we say that a randomly selected number is almost certain to be normal, that is simply a restatement of Borel's normal number theorem in the more-intuitive

language of probability rather than the less-intuitive language of Lebesgue measure.

5. Technically, the temperature sliding down from 70 degrees to 69.9999 degrees is discontinuous unless it passed through every real number between 70 and 69.9999; this is a consequence of the intermediate value theorem for continuous functions. The purpose of this illustration was to give the reader the idea of a nonjumpy transition without getting overly technical.

6. One of these equations is the Navier-Stokes equation, a partial differential equation whose solution is one of the Clay Mathematics Institute's millennium problems.

7. G. J. Sussman and J. Wisdom, "Numerical Evidence That the Motion of Pluto Is Chaotic," *Science* 241: pp. 433–37.

8. See http://www.jaworski.co.uk/m10/10_reviews.html. I can't believe this is still around!

9. For those interested in reading further about the history and early development of chaos, I recommend James Gleick's *Chaos: Making a New Science* (New York: Viking, 1987). Gleick is a terrific science writer, a worthy heir to Paul de Kruif, Isaac Asimov, and Carl Sagan. Chaos has advanced considerably since this book was published, though.

The Raw Materials

The Importance of Being Earnest

In early 1996, the journal *Social Text* published an article by Professor Alan Sokal of New York University. Entitled "Transgressing the Boundaries: Towards a Transformative Hermeneutics of Quantum Gravity"[1] (huh?), the article put forth the viewpoint that "physical 'reality' . . . is at bottom a social and linguistic construct" (say what?). The article was in fact a giant intellectual hoax, and would soon become a cause célebre. Sokal submitted his article because he feared that the view that the world is how we perceive it, rather than how it is, was distorting one of the fundamental goals of science: the search for truth. The acceptance of the article by the journal had numerous side effects. It helped to increase the degree of scrutiny with which articles dealing with technical subjects were examined and also revealed how publication likelihood was affected, at least in the liberal arts,[2] by the concordance of the article with the philosophical or political positions of the editorial staff.

Mostly, however, it helped expose a disturbing trend: the belief that it is the perception of reality, rather than reality itself, that matters most.

Sokal found such a view abhorrent—as would most scientists, whose job it is to investigate reality. As he put it, "I'm a stodgy old scientist who believes, naively, that there exists an external world, that there exist objective truths about that world, and that my job is to discover some of them."[3]

Failure to pay attention to the realities of the external world has been the cause of numerous tragedies, from Icarus to *Challenger*. As Richard Feynman remarked during the investigation that followed the disaster that occurred when the shuttle *Challenger* was launched under unsafe conditions, "For a successful technology, reality must take precedence over public relations, for Nature cannot be fooled."[4]

Among the great objective truths that we have learned about the external world is that not all things are possible. One plus one will always equal two, no matter whether we give 110 percent or resort to wishing on a star[5]—because, at bottom, arithmetic is *not* a social or linguistic construct. If we add up the numbers in our checkbook correctly and the balance is $843.76, that's what we've got. Unfortunately, never more, if we want to buy a Lexus without resorting to five years of monthly payments, and fortunately, never less, if we want to go to dinner and a movie without worrying that we will be thrown in debtor's prison if we can't pay for it.

Nature supplies us with the raw materials from which the universe is constructed. Some of those raw materials are, well, material: the matter from which every thing in the universe is made. Some of those raw materials are less substantial, such as energy. There are relationships between and among the raw materials of the universe that dictate what is and what is not possible. This was first glimpsed by the French chemist Antoine-Laurent Lavoisier, who discovered that in a chemical reaction, the total mass of the products of the chemical reaction was equal to the total mass of the substances that reacted. This result, known as the law of conservation of mass, marked the beginning of theoretical chemistry. A trio of nineteenth-century scientists would significantly expand upon this result, extending to energy what Lavoisier did for matter.

The Heat Is On

In the summer of 1847, William Thomson, a young Briton, was vacationing in the Alps. On a walk one day from Chamonix to Mont Blanc, he encountered a couple so eccentric they could only be British—a man carrying an enormous thermometer, accompanied by a woman in a carriage. Thomson, who was later to become one of the greatest of British scientists and be granted the title Lord Kelvin, engaged the pair in conversation. The man was James Prescott Joule, the woman his wife, and they were in the Alps on their honeymoon. Joule had devoted a substantial portion of his life to establishing the fact that, when water fell 778 feet, its temperature rose 1 degree Fahrenheit. Britain, however, is notoriously deficient in waterfalls, and now that Joule was in the Alps, he certainly did not intend to let a little thing like a honeymoon stand between him and scientific truth.

A new viewpoint had arisen in physics during the early portion of the nineteenth century: the idea that all forms of energy were convertible into one another. Mechanical energy, chemical energy, and heat energy were not different entities, but different manifestations of the phenomenon of energy. James Joule, a brewer by trade, devoted himself to the establishment of the equivalence between mechanical work and heat energy. These experiments involved very small temperature differences and were not spectacular, and Joule's results were originally rejected, both by journals and the Royal Society. He finally managed to get them published in a Manchester newspaper for which his brother was the music critic. Joule's results led to the first law of thermodynamics, which states that energy cannot be created nor destroyed, but only changed from one form to another.

Some twenty years before Joule, a French military engineer named Nicolas Carnot had been interested in improving the efficiency of steam engines. The steam engine developed by James Watt was efficient, as steam engines went, but nonetheless still wasted about 95 percent of the heat used in running the engine. Carnot investigated this phenomenon and discovered a truly unexpected result: it would be impossible to devise a perfectly efficient engine, and the maximum efficiency was a simple mathematical expression of the temperatures involved in running the engine. This was Carnot's only publication, and it remained buried until it was resurrected a quarter of a century later by William Thomson (Lord Kelvin), just one year after his chance meeting with Joule in the Swiss Alps.

Carnot's work was the foundation of the second law of thermodynamics. This law exists in several forms, one of which is Carnot's statement concerning the maximum theoretical efficiency of engines. Another formulation of the second law, due to Rudolf Clausius, can be understood in terms of entropy, a thermodynamic concept that involves a natural direction of thermodynamic processes: a cube of ice placed in a glass of hot water will melt and lower the temperature of the water, but a glass of warm water will never spontaneously separate into hot water and ice.

The Austrian physicist Ludwig Boltzmann discovered an altogether different formulation of the second law of thermodynamics in terms of probability: systems are more likely to proceed from ordered to disordered states, simply because there are a lot more disordered states than ordered ones. The second law of thermodynamics explains why clean rooms left unattended become dirty, but dirty rooms left unattended don't become clean: there are many more ways for a room to be dirty than for it to be clean. The first and second laws of thermodynamics seem to appear in so many diverse environments that they have become part of our collective understanding of life: the first law says you can't win, and the second law says that it's not possible to break even.

Carnot, Joule, and Boltzmann came at thermodynamics from three different directions: the practical (Carnot), the experimental (Joule), and the theoretical (Boltzmann). They were linked not only by their interest in thermodynamics, but by difficult situations bordering on the tragic. Carnot died of cholera when he was only thirty-six years old. Joule suffered from poor health and a childhood spinal injury all his life and, though the son of a wealthy brewer, became impoverished in his later years. Boltzmann was a manic depressive who committed suicide because he feared his theories would never be accepted; ironically, his work was recognized and acclaimed shortly after his death.

The Ultimate Resource

There are striking parallels between energy and money. Each is the ultimate resource in its own particular arena. Money is how we evaluate and pay for goods and services, and energy is the measure of how much effort is necessary to produce those goods and supply those services. Just as different currencies can be exchanged for each other, various forms of energy can be converted into each other.

The first law of thermodynamics, as described earlier, states that there are no free lunches in the universe—energy cannot be created from nothing. Nor, and this is often ignored, can energy be destroyed, but it can be transmuted. The second law, which addresses the transmutation of energy, also has a monetary analogue: in real life, money is never used with perfect efficiency. There are always middlemen extracting money for making arrangements, and nature does the same thing whenever energy is used. Energy can never be used with perfect efficiency; this is one of the reasons that perpetual motion machines can never be built.

Recent developments, though, have made it appear that there may be chinks in the laws of thermodynamics. One such chink is a consequence of a topic discussed in an earlier chapter: only gravity (of the four forces) is capable of exerting an extra-dimensional influence. We can never directly observe a fourth spatial dimension, because the process of observing a fourth dimension involves the use of the electromagnetic spectrum, and current theories do not allow the electromagnetic force to probe a fourth dimension. However, the gravitational force can leak into other dimensions—as we mentioned, this is one way that we might be able to discern the existence of those other spatial dimensions.[6] If this does indeed prove to be the case, the first law of thermodynamics would no longer hold; but it would open up an extremely appealing possibility. If gravitational energy from our three dimensions could leak out elsewhere, why couldn't gravitational energy from other dimensions leak into ours? This might enable us to obtain free lunches from extra-dimensional caterers, and at the same time necessitate a new first law of thermodynamics: in the universe as a whole, energy cannot be created or destroyed. There have been other instances in which the Law of Conservation of Energy has been restructured. Einstein's classic equation $E = mc^2$ gives the "exchange rate" for matter and energy; 1 unit of matter is converted into c^2 units of energy. This necessitated a restatement of the law of conservation of energy: the totality of matter and energy are conserved according to Einstein's formula, much as the total value of cash remains the same even if some of it is in dollars and some in euros. Given this history for the law of conservation of energy, it would not be completely surprising if yet another change lurked in its future.

Why Entropy Increases

In order to know why entropy increases, we have to know how to calculate entropy. The symbol Δx, which appears frequently in mathematics, represents the change in the quantity x—if, at the end of the month, x represents my bank balance, Δx is the amount of money that the state of California, which employs me, deposits directly to that account. In thermodynamics, S represents the amount of entropy in the system. The symbol ΔS, the change in the entropy of the system, is the sum of all the quantities $\Delta Q/T$ in a system, where T is a temperature at which a component of the system resides and ΔQ is the heat change in that component at the temperature T. For those who have had calculus, it's more formally defined as the integral of dQ/T—for those who haven't had calculus, an integral is simply the sum of lots of very small things.

I'll borrow an example from Brian Greene's *The Fabric of the Cosmos*[7] and imagine that we have a glass of water with some ice cubes in it. Heat flows from hotter to colder because heat is a measure of how fast molecules are moving; when fast-moving molecules collide with slow-moving ones, the faster ones slow down (losing heat) and the slower ones speed up (gaining heat). Let's assume that 1 unit of heat is transferred from a small amount of water at temperature T_1 to an ice cube at temperature T_2. Since water is warmer than ice, $T_2 < T_1$.

Heat accounting is very similar to checkbook accounting; units gained are viewed as positive (we add deposits in our checkbook), and units lost are viewed as negative (we subtract checks or withdrawals, and subtracting a positive number yields the same result as adding a negative one). So the contribution to the change in entropy from the loss of the heat unit from the small amount of water is $-1/T_1$. The contribution to the change in entropy from the gain of the heat unit by the ice cube is $+1/T_2$. The total change in entropy from this heat transaction is $-1/T_1+1/T_2$; this expression is positive since $T_2 < T_1$. As the water cools and the ice melts, each one of these heat transactions changes the entropy by a positive amount, and so the entropy of the system increases.

Once the system has reached equilibrium, with all the cubes melted and the system at a uniform temperature, no more heat transactions can take place and the glass of water is at maximum entropy. The glass of water is a microcosm of what is happening in the universe. For the most part, warm things are cooling and cool things are warming, entropy is increasing, and we are headed toward a dim and distant future where everything is at the same temperature, no more heat transactions can take place, and things are really, really dull because nothing can happen. This is the so-called heat death of the universe.

At least entropy doesn't always increase everywhere at every time; the second law only requires entropy to increase in reversible procedures, and, fortunately, a lot of the really interesting procedures do not fall into that category. The freezing of ice cubes, or the birth of a child, requires a local decrease in entropy—but it is always at the expense of the increase in entropy in the universe as a whole, because the universe must supply heat to run the refrigerator to freeze the ice cubes, and in order to produce the child it requires a lot of entropy in the form of material and energy.

Local decreases in entropy take place for a variety of reasons, not just because you need to use electricity to run your refrigerator to make the ice cubes. Gravity, of which there's a lot lurking around the universe, contributes to local decreases in entropy that help power our existence. A cloud of hydrogen gas, when viewed strictly from the standpoint of its thermodynamic properties, is a high-entropy system. What the thermodynamic viewpoint fails to take into consideration is the role that gravity plays in causing a local entropy decrease. The cloud, if it is large enough, collapses under its own gravitation until its mass is dense enough to cause thermonuclear fusion, and a star is born. If the star is large enough, an even more dramatic entropy decrease is in the offing, as the star will eventually explode in a supernova, a process that creates the heavy elements from which planets, and living things, can eventually form.

Another Look at Entropy

Statistical mechanics offers an alternative definition of entropy. Statistical mechanics arose from the problem of discovering and utilizing the vast amount of information there is in any assemblage of molecules. Any sizable assemblage of molecules, such as a glass of water, contains at least 10^{24} molecules, each occupying a specific location (requiring three coordinates to specify) and moving in three different directions (also requiring three coordinates to specify the north-south velocity, the east-west velocity, and the up-down velocity). Even if we could acquire knowledge (which we can't) of all this information for every molecule in the glass of water, what on Earth would we do with it? Talk about information overload! If each computer had a terabyte of storage (a trillion bytes; I wouldn't be surprised if they're on the market soon, if they're not already) and each coordinate used a single byte, you would need a computer for every man, woman, and child on Earth simply to store that information about a glass of water.

We encounter the same problem in analyzing the attributes of large assemblages of anything, such as the income distribution of the population of the United States. The IRS undoubtedly has reasonably accurate date for, say, 100 million people, but if we had a book with the information for all 100 million people in it, our eyes would undoubtedly glaze as we tried to examine it. Boil it all down to a small chart, such as the percentage of people making less than $25,000, the percentage making between $25,000 and $50,000, the percentage making between $50,000 and $75,000, the percentage making between $75,000 and $100,000, and the percentage making more than $100,000, and we are much more able to appreciate it and use it to make decisions. Statistical mechanics was born when it was realized that similar principles applied to the positions and motions of large assemblages of molecules.

Any macrostate of a system, such as a glass of water with ice cubes in it, is an assemblage of microstates—the temperature, velocity, and location of the individual molecules. The definition of entropy offered by statistical mechanics is a measure of the number of microstates associated with each macrostate. A glass of water with ice cubes in it has fewer microstates comprising it than a glass of water at a uniform temperature, because we are confining the ice cube molecules substantially. Their location and velocity is highly restricted, whereas an individual water molecule is free to zip around and go anywhere. The second law of thermodynamics, in this viewpoint, is a statement about probability; it is more likely that if a system evolves from one state to another, it will move toward a state with higher probability. When we throw a die, it is less likely to land less than 3 than greater than 3 because there are only two states less than 3 (1 and 2) but three states greater than 3 (4, 5, and 6).

This gives a probabilistic explanation of why the ice cubes melt: there are fewer states with ice cubes and hot water than with lukewarm water at a uniform temperature. It also points out why systems tend toward equilibrium: these are the states of highest probability, and any deviation therefrom will naturally tend to evolve back toward a state of higher probability.

However, the statistical view of the second law opens a door that is hidden in the classical formulation. A system is not compelled to be in its maximum-probability state, it is just more likely to be there than anywhere else. Unlikely though it may be, a glass of water at uniform temperature may undergo a highly unlikely series of transitions, resulting in a glass with rectangular cubes of ice immersed in hot water—or, even more unlikely, with ice cubes shaped like miniature replicas of the Parthenon. When I was younger, I read *One, Two, Three . . . Infinity*, a wonderful book by the physicist George Gamow.[8] In it, he describes a similar situation in which all the air molecules in a room migrate to an upper corner, leaving the unfortunate inhabitants gasping for breath. He then does the calculation, and shows that we would have to wait roughly forever for this to happen. I confess that I was definitely relieved to hear that this was one more thing I wouldn't have to worry about—but I probably wouldn't have worried about it had I not read the book.

Order and Disorder

Everyday life offers us a way to visualize the number of microstates associated with a given macrostate. Linda, my wife, and I have a very different view of the purpose of my closet. Linda thinks that the closet exists for clothes to be hung in their proper places. Left to her own devices, she sorts the hangers from left to right; shirts on the left, pants on the right. The shirts and pants are further subdivided by whether they are work clothes (characterized by whether they have stains from the colored overhead markers I use when lecturing) or dress clothes (those that are still pristine; all shirts and pants were initially purchased to be dress clothes, but Linda realizes that they are not all destined to remain so). A final order is imposed when she sorts them by color. I haven't figured out what her color sorting strategy is; if somebody compelled me to do it, I would probably go with ROY G BIV (an acronym long ago committed to memory for the colors of the spectrum in the order they appear in a rainbow: red, orange, yellow, green, blue, indigo, violet). However, I have a friend who sorts his books by color and does so alphabetically; he would sort the above colors in the order blue, green, indigo, orange, red, violet, yellow).

I, on the other hand, have a total disregard for such niceties. If the clothes are on hangers, it's fine with me. So it takes me a little extra time to find the right shirt and pants. Big deal. Linda scrutinizes the closet every few months, and her reaction is always the same: I've messed things up again. There is only one right order in which to hang the clothes, and all other orders are characterized by the phrase "messed up." For her, there are only two macrostates: the correct order (and there's only one microstate corresponding to the "correct order" macrostate) and messed up (and she believes that I've generated every possible microstate corresponding to the "messed-up" macrostate).

Nature and my closet have this in common: there are considerably more microstates corresponding to a disordered macrostate than to an ordered one. For a quantitative demonstration of this, suppose that we have two pairs of shoes: sneakers (what I wear most of the time) and loafers (what I wear on formal occasions), and two boxes in which to put the shoes. These boxes are large enough for all four shoes to fit, but one box is for the sneakers, and one for the loafers. The following is a table of all the different ways to put the four shoes in the two boxes. There is an obvious quantitative criterion for what constitutes order in this situation: the number of pairs of shoes in the proper box. The abbreviation LB stands for loafer box, and SB is for sneaker box.

Left Loafer	Right Loafer	Left Sneaker	Right Sneaker	# Pairs in Correct Box
LB	LB	LB	LB	1
LB	LB	LB	SB	1
LB	LB	SB	LB	1
LB	LB	SB	SB	2
LB	SB	LB	LB	0
LB	SB	LB	SB	0
LB	SB	SB	LB	0

Left Loafer	Right Loafer	Left Sneaker	Right Sneaker	# Pairs in Correct Box
LB	SB	SB	SB	1
SB	LB	LB	LB	0
SB	LB	LB	SB	0
SB	LB	SB	LB	0
SB	LB	SB	SB	1
SB	SB	LB	LB	0
SB	SB	LB	SB	0
SB	SB	SB	LB	0
SB	SB	SB	SB	1

There are three macrostates: 2 pairs in the correct box (the most ordered macrostate), 1 pair in the correct box, and 0 pairs in the correct box (the most disordered macrostate). There is 1 microstate corresponding to the most ordered macrostate, 6 microstates corresponding to the next-most-ordered macrostate, and 9 microstates corresponding to the most disordered macrostate. It doesn't always work out as neatly as this, but the more the possible number of microstates, the more likely it is that the ordered macrostates are far less probable than the disordered ones. What makes it so unlikely that the air molecules in the room will all migrate to the upper three inches is that there are on the order of 10^{25} air molecules in the room, and the number of microstates in which all these molecules are up near the ceiling is almost infinitesimal when compared with the number of microstates in which the molecules are spread out all over the room (to the great relief of the inhabitants of that room).

Entropy and Information

We are living in what has been described as the Information Age. America, the country whose vast industrial complex once churned out the automobiles and refrigerators that contributed so greatly to its wealth, has nearly abandoned the production of these material commodities to places where it can be done more efficiently (thanks to more modern production equipment) or more cheaply (thanks to an abundant supply of labor). Yet America still retains much of its leadership of the industrial world because the new coin of the realm is neither automobiles nor refrigerators, but information; and America is at the forefront of the manufacture and distribution of information.

But how does this relate to the concepts we have been examining? It was

Boltzmann, the architect of statistical mechanics, who realized that when one talks about order and disorder, about the number of microstates associated with a given macrostate, the concepts being discussed related to the information one had about the system. Let's look again at the example of the loafers and sneakers above. The most ordered macrostate is the one where every shoe is in its correct box, the least ordered one is the one where no pair of shoes is in its correct box, and the number of microstates corresponding to each varies inversely with the precision with which we can locate the shoes if all we know is the macrostate.

The more microstates that are associated with a given macrostate, the less we are able to say with precision about the individual components of the system. When we know that both pairs of shoes are in the correct box, we know with certainty where the left loafer is. By checking the table on pages 193–194, if we know that only one pair of shoes is in the correct box, the left loafer is in the loafer box in 4 out of 6 microstates—a probability of 2/3. However, if no pair of shoes is in the correct box, the left loafer is in the loafer box in only 3 of the 9 microstates, a probability of 1/3. This analysis is typical; as entropy increases, the information that we have about the system decreases. Since the second law of thermodynamics tells us that entropy is on the rise in the universe as a whole, the inexorable progress of time is increasing what we cannot know. The heat death of the universe is also an information death; the universe is tending toward a state in which there is nothing left to do, and very little of a physical nature to know.

This is the exact opposite of what our everyday experience tells us. Every day science gathers more and more information about the universe around us; but that is because entropy is still capable of suffering local defeats. There is still lots of information to be gathered, and there will be into the far distant future. But even as we are greedily sucking up exponentially more information, in the far, far, far distant future we will head inexorably toward a universe in which we cannot know almost anything, because there will be almost nothing to know.

Black Holes, Entropy, and the Death of Information

As science progresses, many of its important ideas traverse a common path. The first stage is the formulation of a hypothetical construct (the Sokal article now has me using that term), an object whose existence explains certain phenomena. The next stage is indirect confirmation; experiments or observations suggest that the construct does indeed exist. Finally, we hit the jackpot, a direct observation of the object under consideration.

This path is traveled not just in the physical sciences, but in the life sciences as well; it describes both the atom and the gene.

It also describes the black hole, whose existence was first hypothesized more than two centuries ago by the English geologist John Michell. In a paper published by the Royal Society, Michell stated, "If the semi-diameter of a sphere of the same density as the Sun were to exceed that of the Sun in the proportion of 500 to 1, a body falling from an infinite height toward it would have acquired at its surface greater velocity than that of light, and consequently supposing light to be attracted by the same force in proportion to its vis inertiae (inertial mass), with other bodies, all light emitted from such a body would be made to return towards it by its own proper gravity."[9] The basic idea of a black hole is clearly contained in this statement: the gravity of the object is so strong that no light can escape from it.

With the development of Einstein's theory of relativity, interest in the concept of a black hole picked up steam. In the 1930s, work was initiated by the astrophysicist Subrahmanyan Chandrasekhar and continued by Robert Oppenheimer (among others), who in just a few years would head the Manhattan Project, which developed the first atomic bomb. They concluded that stars possessing greater than a certain mass would undergo an unstoppable gravitational collapse and become a black hole. Black holes thus progressed from hypothetical construct to entities that might conceivably be observed, either indirectly or directly. Supermassive black holes, with masses millions of times the mass of the Sun, are now believed by many physicists to lurk at the core of major galaxies, including the Milky Way galaxy in which Earth resides. In 2004, astronomers claimed to have detected a black hole orbiting the supermassive black hole at the center of the Milky Way galaxy (fortunately, Earth is situated a comfortable distance away from the center).[10] Although we will never see a black hole, as John Michell was well aware, the evidence for their existence is now extremely strong.

What is known about black holes is that they are completely determined by their mass, their charge, and their spin. These are the only things we can ever know about a black hole, and so when we see a black hole with a given mass, charge, and spin (the macrostate), all the gazillions of possible microstates occurring within the black hole correspond to that single macrostate. Black holes are therefore the ultimate limit of how high the entropy can go. The higher the entropy, the less the information, and a black hole of a given mass, charge, and spin conveys the least possible information about the region of space it occupies. The goings-on in the inside of the black hole appears to be high on the list of things we cannot know.

Over the course of the next decade or so, astrophysical measurements should reveal the future of the universe: whether it is destined to expand forever or whether it will eventually recollapse in what has been termed the big crunch. Black holes will merge with other black holes until there may be only one giant black hole with all the matter in the universe, collapsing ever in on itself.

This was the conventional view of black holes until Stephen Hawking showed in the 1970s that black holes are not as black as initially thought. Quantum-mechanical processes allow matter to escape from the black hole in a process known as Hawking radiation.[11] Just as water in a glass slowly evaporates as its individual molecules acquire enough velocity to escape the bounds of the glass, the matter within a black hole evaporates over time. Surprisingly, though, the rate at which the matter disappears depends strongly on the size of the black hole. A black hole the size of the Sun will take on the order of 10^{67} years to evaporate. Considering that the age of the universe itself is approximately 10^{14} years, solar mass black holes will be hanging around until the far, far, far distant future. Should the universe collapse into a black hole via the big crunch, it may take close to forever for it to evaporate, but evaporate it will.

The Universe and Princess Leia

Hawking's work also led the way to the surprising result that the entropy of a black hole is proportional to its surface area, rather than its volume. What makes this result surprising is that we have already seen that entropy is a measure of disorder, and we would certainly expect volume to be capable of displaying more disorder than the vessel that contains the disorder.

As a result, some physicists have speculated that all the order and disorder we see in our universe is merely a projection of order and disorder on a multidimensional boundary that in some sense encloses our universe the way the surface of a basketball encloses the region inside it. This is something akin to the way a hologram works. A hologram is a clever device that projects the illusion of a three-dimensional object from the information inscribed on a two-dimensional one.

There is a scene early in Episode 4 of *Star Wars* (that's the first one filmed back in the mid-1970s) in which Luke Skywalker and his droids discover an old holographic projection device. They crank it up and a holographic image of Princess Leia appears. The image is a little fuzzy, but it is nonetheless three-dimensional, and the holographic Princess Leia certainly possesses a considerable amount of passion as she makes her pitch for help. We know the holographic image of Princess Leia is just a holographic image, but we can conjecture about whether the image, if somehow it were imbued with consciousness, would realize that it is only that, a holographic projection. If this is the way the universe is, and we are but holographic projections of some sort, how would we ever know?

What Mathematics Has to Say About This

From a mathematical standpoint, it is easy to put lower-dimensional objects into one-to-one correspondence with higher-dimensional objects. To see a simple example of how to put the points on a line segment into one-to-one correspondence with the points in a square, take a number between 0 and 1 and write out its decimal expansion, then simply use the odd-numbered digits (the tenths, thousandths, and hundred thousandths, etc.) to define the x coordinate and the even-numbered digits (the hundredths, ten thousandths, and millionths, etc.) to define the y coordinate. Thus, the number .123456789123456789123456789 . . . would correspond to the point in the square (.13579246813579 . . . , .2468135792468. . . .).

To map points in the square to the line, simply interleave the digits alternately—the exact reverse of what was done above. The point (.111111 . . . , .222222 . . .) would correspond to the point .12121212 . . . on the line.

The problem here is that these transformations are discontinuous. Points close to one another can end up widely separated, analogous to the baker's transformation we saw when we were discussing chaos. In fact, it can be shown in topology that every one-to-one transformation of the portion of the real line between 0 and 1 onto the square in the plane must be discontinuous.[12]

This has interesting consequences for Princess Leia, as well as for us, if we are holographic projections. One would expect that the recipe for constructing the holographic Princess Leia would be continuous in the following sense: just as all the points in Princess Leia are close to one another (at least, they are all within Princess Leia), the recipe for constructing her would consist of instructions close to one another. But mathematics shows this cannot be the case. The recipe for constructing Princess Leia (or, if not Princess Leia, some other holographic projection) must be widely scattered. Instead of a single block of instructions on how to construct Princess Leia, the instructions for doing so may appear all over the book that represents the totality of all the holographic recipes. We might expect that pages 5–19 of the holographic recipes are devoted to construction of

Princess Leia, but what might actually happen is that one line of page 5 might concern itself with the princess, with the next line on Princess Leia appearing on page 8,417,363.

The Blind Holographer

I don't know about you, but this makes me doubt the holographic explanation. I'm not comfortable with the idea that the instructions for making yours truly are scattered all over hell and gone; I'd be a lot more comfortable if they were close together, like me. There is a hole in this argument, though: the theorem from topology that states that any transformation between objects of different dimensions must be discontinuous relies on the fact that the objects are continua of higher dimensions, characterized by the real numbers. The universe is quantized, the elements in it discrete, and this would also be true of the boundary of the universe (I think). In this case, my objections might not be valid, but I've never seen a theorem from topology addressing this point.

Another reason that casts doubt, at least for me, upon the universe-as-hologram theory is an updated version of the watchmaker theory espoused by the eighteenth-century theologian William Paley. Paley argued that just as a watch is too complicated a contrivance to simply occur naturally and the existence of a watch implies the existence of a watchmaker, so do the complexity of living things imply the existence of a creator. In his 1986 book, *The Blind Watchmaker*,[13] Richard Dawkins argued that natural selection plays the role of blind watchmaker, without purpose or forethought, but directing the evolution of living things.

The updated version of Paley's argument would be that the existence of a hologram (the universe) implies the existence of a holographer outside the universe. Just as a character in a novel cannot write the novel in which he is a character, a character in a hologram cannot create the hologram. So where does the hologram come from, and where do the rules come from that govern how the hologram is played? Maybe it's like a self-extracting file on a computer. I confess that my knowledge of the physics behind this theory is nonexistent, but nonetheless it seems to me that the universe-as-hologram theory requires either a blind holographer, a concept that is to this theory what natural selection is to evolution, or the existence of a holographer outside the universe. Possibly the blind holographer will prove to be as much of a bombshell for physics as natural selection was to evolutionary biology.

NOTES

1. See http://skepdic.com/sokal.html. This is an excellent summary of the Sokal hoax. *The Skeptic's Dictionary* has a lot of good stuff, especially for those of us who are confirmed skeptics. It has great sections on UFOs, the paranormal, and junk science. You can get more of an education from this site than you can from a degree in practically any one of the currently trendy areas of academia.

2. It pains me to say this, but the hard sciences and mathematics are not immune from this. People have a difficult time with ideas that challenge their cherished beliefs. That's why I'm such a great respecter of science; it has a built-in mechanism (replicability) to counter this.

    This also works when something challenges the established paradigm. Cold fusion sounded great, but when nobody could duplicate the critical experiments, it passed from view.

3. See http://skepdic.com/sokal.html.

4. See http://www.brainyquote.com/quotes/authors/r/richard_p_feynman.html. The quotes on this site are well worth the five minutes it takes to read them. Feynman died before the Sokal hoax, but the following quote is certainly applicable: "The theoretical broadening which comes from having many humanities subjects on the campus is offset by the general dopiness of the people who study these things."

5. The song lyrics "When you wish upon a star / Makes no difference who you are" nails it, because nothing will happen that wouldn't have happened anyway. Remarks like this flow freely when you spend a little time with *The Skeptic's Dictionary*.

6. So far, no such leakage has been detected. That doesn't mean that reasonable theories based upon such a leakage can't be constructed, although they may be hard to test. Recall that the steady state theory required the creation of one hydrogen atom per cubic meter every 10 billion years. The steady state theory did make predictions (or rather, it didn't make the key predictions that the big bang theory did), and as a result it was possible to reject it based on experimental evidence.

7. B. Greene, *The Fabric of the Cosmos* (New York: Vintage, 2004), pp. 164–67. I've said this before: this is a terrific book. Not an easy book (don't believe blurbs to the contrary), but utterly, completely, and totally worth the effort.

8. G. Gamow, *One, Two, Three . . . Infinity* (New York: Viking, 1947). This was the book that started me on math and science, and if you have bright, inquisitive children age twelve or older, give this to them. There will be a bunch of math they won't understand, but a lot that they will, and some of the science is obsolete or erroneous, but who cares? That can be corrected, and none of the math is erroneous.

9. See http://www.manhattanrarebooks-science.com/black_hole.htm. This quote can be found in more erudite sources, I'm sure.

10. See http://www.mpe.mpg.de/ir/GC/index.php. Nice photos and graphics from the Max Planck Institute.

11. See http://en.wikipedia.org/wiki/Hawking_Radiation. There's probably more math here than you want to absorb, but this is a good article explicating the basic ideas.

12. There are many different ways of showing this fact (in fact, when I give exams in a first-year course in topology, I ask students to show this in at least two ways). One way to show that the unit interval cannot be mapped continuously onto a solid square one unit on a side is to take one point out of the middle of each. Doing so separates the unit interval into two distinct pieces; in the language of topology, it is disconnected. However, removing a point from the middle of the solid unit square still leaves an object that is connected; you can walk from any point to another with your path still in the square without the point, like avoiding a gopher hole in your back yard.

13. R. Dawkins, *The Blind Watchmaker* (New York: W. W. Norton, 1986). This is a great book, but it touches on ideas that make some people nervous. Dawkins has since become one of the leading spokesmen for atheism. Of course, that's his point—that evolution can occur without the direction of a creator. I think everyone should read this book, as it will make you think, no matter what your point of view.

Section IV

The Unattainable Utopia

12
Cracks in the Foundation

The Foundation of Democracy

Elections are the foundation of democracy. We vote on matters as important as who will be the next president, and as trivial as who will be the next American Idol.

When a candidate receives a majority of votes, there is no difficulty determining the winner of the election—but when no candidate receives a majority of the votes, problems can arise. Although the winner of the election is sometimes unclear (as in the 2000 presidential election), often the election is decided by the choice of the rules that govern the election. American elections have a long history of exposing the soft underbelly of the election rules that are currently in use.

Presidential elections currently use the Electoral College rather than the popular vote to decide the outcome, but this system first ran into trouble in the election of 1800, when Thomas Jefferson and Aaron Burr, the two leading candidates, received the same number of electoral votes. The Twelfth Amendment was passed to resolve the problems that occurred in this election by transferring the election to the House of Representatives in case no candidate has an electoral majority, but the 1824 election revealed that the system had not been fixed. The leading candidate, the son of a former president of the United States, failed to win the popular vote. The election was eventually decided not by the voters but by a relative handful of highly placed government officials. It may sound like a description of the 2000 presidential election, but history does have a tendency to repeat itself.

The 1824 election featured four major candidates: the charismatic general Andrew Jackson, who had helped defeat the British in the War of 1812; John Quincy Adams, the son of a former president who was the secretary of state at the time of the election; William Crawford, the secretary of the treasury; and Henry Clay, the Speaker of the House. After the votes were cast, Jackson had received a plurality of both the popular vote and the Electoral College vote, but he had not obtained the needed Electoral College majority. As provided by the Twelfth Amendment, the election went to the House of Representatives (a specter that briefly reappeared during the 2000 election as well); but the Twelfth Amendment stipulated that only the top three vote getters could be considered. This eliminated Clay, who encouraged his electors to vote for Adams, a man he disliked personally but with whom he shared some important political views. As a result, Adams won, despite the fact that Jackson had received not only the most popular votes, but also the most votes in the Electoral College. When Adams later appointed Clay as secretary of state, it seemed to many that this was the payoff for Clay's votes.

Even today, the system still has not been fixed. The Electoral College places a different weight on popular votes cast in different states, and trying to assess the relative weights of those votes is not an easy task. If one defines the value of an individual's vote as the fraction of an electoral vote that it represents, then votes of individuals in low-population states with three electoral votes are often considerably more valuable than the votes of individuals in populous states such as California or New York. Using this method of evaluation, a Wyoming voter has almost four times the Electoral College clout as a California voter.[1]

There is an alternative—and more mathematically interesting—way to measure the weight of a vote. The Banzhaf Power Index (BPI) counts how many coalitions (a coalition is a collection of votes) a voting entity can join such that its joining that coalition changes the coalition from a losing one to a winning one.

Although the BPI can be computed for both individual voters and blocs of voters, it is easier to understand how it is computed in the context of an electoral college. To see how the BPI is computed, suppose there are three states in an electoral college of 100 votes. The states have 49, 48, and 3 electoral votes. To compute the BPI of the 3-vote state, we count the losing coalitions that become winning ones once the 3-electoral-vote state joins it. By itself, the 49-vote state is a losing coalition, but if the 3-vote state joins it, the 52-vote total assures victory. Similarly, by itself the 48-vote state loses, but if the 3-vote state joins it, the coalition is a winning one. This computation shows that the 3-vote state has a BPI of 2, as does each of the other states. The small state has tremendous clout in this election, far out of proportion to its actual electoral total, and the candidates should be working just as hard to win this state as either of the big ones.

The opposite side of this picture is that a state with an apparently sizable number of electoral votes may actually be powerless to swing an election. If there are three states, each with 26 electoral votes, and a fourth state with 22 electoral votes (again, the electoral vote total is 100), whichever candidate wins two of the large states wins the election; it makes no difference how the small state votes. There is no losing combination of states the small state can join that will turn that combination into a winner; the small state has a BPI of 0. Each of the big states has a BPI of 4, as it can join either of the other big states, or an alliance of a big state plus the small state, and turn a losing combination into a winning one. The voters of the small state are effectively disenfranchised.

Power index analyses have been done that show that a California voter is almost three times as likely to swing a presidential election as a voter from the District of Columbia.[2] So everyone knows that the Electoral College is not a truly democratic way to decide a presidential election—but how we measure this depends upon the mathematics one chooses to use to analyze the situation.

In an ideal democracy, each vote should have equal weight, so we might decide to give each voter a total of 100 points, and ask him to distribute those points among the various candidates. Called the preference intensity method, a variation of this was used for more than a century in determining the members of the Illinois House of Representatives.[3] A simple example shows that there are potential problems with this method. Consider an election with two candidates, A and B, and three voters (we could as easily be discussing larger groups of people as well as individuals). The first voter allocates 100 points to A and none to B, whereas the other two voters allocate 70 points to B and 30 points to A. A majority prefers B to A, but the lone voter that prefers A to B carries the day, as A wins the election by 160 points to 140. Are we content with an election process that allows a vocal minority to outshout the majority?

The investigation of the problems in determining which voting method to use in a democracy goes back more than two centuries. Probably the first person to notice that there were problems associated with determining the preferences of the majority was a French mathematician-turned-bureaucrat who played a significant role in the French Revolution.

A Voting Paradox

Marie-Jean-Antoine-Nicolas de Caritat, Marquis of Condorcet, was born in 1743, at a time when it was a very good thing to be a marquis. Among the numerous benefits conferred upon members of the aristocracy was the availability of higher education. While at college, he focused on math and science, and at graduation was well on the way to becoming one of the leading mathematicians of the eighteenth century. Joseph-Louis Lagrange, a brilliant mathematician and physicist who did groundbreaking work in the theory of probability, differential equations, and orbital mechanics (Lagrange points are locations at which small bodies orbiting two larger ones appear not to move), described Condorcet's thesis as being "filled with sublime and fruitful ideas which could have furnished material for several works."[4] Praise from Lagrange was praise, indeed.

However, soon after the publication of this paper, Condorcet met Anne-Robert-Jacques Turgot, an economist who later became controller general of Finance under Louis XVI. The friendship blossomed, and Turgot arranged for Condorcet to be appointed inspector general of the mint, a similar position to the one awarded Isaac Newton by the English government.

When the French Revolution began, it was considerably less than a good thing to be a member of the aristocracy, but Condorcet actively welcomed the forming of the new Republic. He became the Paris representative to the Legislative Assembly, then later the secretary of the assembly, and helped to construct a plan for a state education system. Unfortunately for Condorcet, when the French Revolution underwent a sea change, he made two critical mistakes. He could possibly have survived his first mistake, which was joining the moderate Girondists and asking that the king's life be spared. His second mistake, though, proved to be fatal. Condorcet failed to recognize that control of the Revolution was about to be seized by the more radical Jacobins. He argued vigorously for a more moderate constitution, which he had helped to write, and soon found himself on the French Revolution equivalent of an enemies list. A warrant for his arrest was issued, and Condorcet went into hiding. He later attempted to flee, but was caught and sent to prison. Two days later, in 1794, he was found dead in his cell. It is not known whether he died of natural causes or was killed.

Condorcet's present fame rests neither on his mathematical discoveries nor his role in the French Revolution, but is based to a much greater extent on what is known as the Condorcet paradox. This may be somewhat of a misnomer, as it is more of an eyebrow raiser than an actual paradox. The Condorcet paradox was the first difficulty discovered in the quest for the ideal voting system. It can occur when the ballot contains three or more candidates, and the voters are asked to rank them, from top to bottom, in order of preference. Rank-order voting has been adopted for national elections by several countries (Australia is a leading example) and although it is not used in the United States in national elections, it is used in some local elections, and is gaining ground. There are at least two good reasons to consider rank-order voting: it helps us prioritize our options, and it does a much better job of eliminating the need for runoff elections (which are both time consuming and expensive) than does simply voting for a single candidate.

One of the chief problems of a society is how we should allocate our resources. Three items of current concern to which we must allocate resources are terrorism, health care, and education. To illustrate the Condorcet paradox, suppose we polled three different individuals to rank these items in order of importance. Here are the ballots that were collected.

	First Choice	Second Choice	Third Choice
Ballot 1	Terrorism	Health care	Education
Ballot 2	Health care	Education	Terrorism
Ballot 3	Education	Terrorism	Health care

Two out of three voters felt that terrorism was more important than health care, and two out of three voters felt that health care was more important than education. If an individual voter felt that terrorism was more important than health care, and health care was more important than education, that voter would logically feel that terrorism was more important than education. But the majority does not appear to behave so logically; two out of three voters feel that education is more important than terrorism! If we are using the majority decision to determine how to allocate funds, we run into an insurmountable problem: it is impossible to spend more on terrorism than on health care, more on health care than on education—and more on education than on terrorism.

This simple example illustrates a problem that would perplex social scientists for more than a century: How can one translate a collection of individual ballots, consisting of a ranking of preferences, into a ranking of preferences for the group as a whole? The Condorcet paradox points out that if we simply look at pairs of candidates and determine which of the pair the majority prefers, we run into the problem that it is impossible to maintain what mathematicians call transitivity, which is a property of relationships such that if A is preferred to B and B is preferred to C, then A is . preferred to C. Individual preferences are transitive; it seems reasonable to require that whatever method we adopt to discover group preferences, what we discover should be transitive as well. The Condorcet paradox highlights the need to determine precisely what properties we want in going from a collection of individual rankings to a societal ranking.

So Who *Really* Won?

As Western civilization began its gradual march toward democracy, different methods of translating individual rankings to societal ones were suggested. It soon became apparent that the outcome of elections could depend on what voting method was adopted.

Numerous methods have been used to determine the winner of an election in which rank-order voting is used. Some of the methods that have been frequently employed to determine a winner are

1. *Most first-place votes.* The winning candidate is the one who receives the most first-place votes. This method, which has been described as winner take all, is used in England to elect members of Parliament, and is common in the United States as well on all levels.
2. *Runoff between the top-two first-place vote getters.* The candidates who place either first or second in the number of first-place votes are matched head-to-head, and the winner is the one preferred by a majority of voters. If voters submit preference lists on the ballot, this method does not need a second election (which is customarily held in the real world, and results in additional monetary costs to both candidates and governments), as it is easy to compute which of the two candidates is preferred by a majority of voters. Nevertheless, top-two runoffs are used in electing the mayors of many large cities, including New York, Chicago, and Philadelphia.
3. *Survivor* (after the popular television show). The candidate who receives the fewest first-place votes is voted off the island. That candidate is removed from consideration and stricken off the existing

ballots, which are then reexamined. The whole process is repeated, until only two remain. The winner of this two-person contest is the one preferred by the majority. The fact that it is possible to avoid run-offs with this method has resulted in it being referred to as instant runoff voting (IRV). This method is used by Australia in its national elections, and was adopted in Oakland, California, in 2006.

4. *Numerical total.* Each voter ranks the candidates on a ballot, and each rank is assigned a point value: for example, a first-place vote could give a candidate 5 points, a second-place vote 4 points, and so on. The winner is the candidate who receives the most total points. This method was suggested by the fifteenth-century mathematician Nicholas of Cusa for electing the Holy Roman Emperors,[5] but today it is used only for major elections in a few small countries. However, it is widely used in nonpolitical elections; the American and National League most valuable players are chosen this way.

5. *Head-to-head matchups.* Each candidate is matched head-to-head with each other candidate. The candidate with the most number of head-to-head victories is the winner. One major advantage of this method, as we shall see later, is that it avoids the Condorcet paradox. However, this method will frequently not give a clear winner; if two candidates tie for first using this method, the winner is determined by the result of the head-to-head matchup between these two candidates. The most widespread use of this method is in round-robin tournaments, which frequently occur in sports, or in games such as chess.

Each of these methods has its advocates, and each has been used in many elections. However, the hypothetical election shown in the table gives us a sense of how difficult it may be to find a good method of selecting a winner from a collection of preference rankings. Five candidates— A, B, C, D, and E—are running for office. Fifty-five ballots were cast, but only six different preference rankings occurred. Here are the results.[6]

Number of Ballots	First Choice	Second Choice	Third Choice	Fourth Choice	Fifth Choice
18	A	D	E	C	B
12	B	E	D	C	A
10	C	B	E	D	A
9	D	C	E	B	A
4	E	B	D	C	A
2	E	C	D	B	A

Admittedly, the above tabulation does seem to resemble something you see when you visit your optometrist, but let's check the results of each method of voting.

1. *Most first-place votes.* A is a clear winner.
2. *Runoff between the top two.* A and B are the two candidates in the run-off, as they have 18 and 12 first-place votes, respectively. A is preferred to B by only the 18 voters who placed him first; B is preferred to A by the other 37 voters, and so B is the winner.
3. *Survivor.* This takes a little work to determine the winner. E receives the fewest first-place votes in the first round, and is therefore eliminated. The table now looks like this.

Number of Ballots	First Choice	Second Choice	Third Choice	Fourth Choice
18	A	D	C	B
12	B	D	C	A
10	C	B	D	A
9	D	C	B	A
4	B	D	C	A
2	C	D	B	A

It's the end of the line for D, who received only 9 first-place votes. The table now reduces to this.

Number of Ballots	First Choice	Second Choice	Third Choice
18	A	C	B
12	B	C	A
10	C	B	A
9	C	B	A
4	B	C	A
2	C	B	A

At least the table is getting easier to read. B receives only 16 first-place votes, leaving A and C in a two-person race. Eliminating B leaves this.

Number of Ballots	First Choice	Second Choice
18	A	C
12	C	A
10	C	A
9	C	A
4	C	A
2	C	A

So C survives by a vote of 37 to 18.

4. Numerical total. We'll assume that the rule here is 5 points for a first-place vote, 4 points for a second-place vote, and so on. You don't have to spend time hauling out the calculator. I'll do it for you. The result is that D wins this contest, with $191 = 5 \times 9 + 4 \times 18 + 3 \times (12 + 4 + 2) + 2 \times 10$ points.

5. *Head-to-head matchups.* You can probably see this one coming. E wins all head-to-head matchups. He or she beats A by 37 to 18, B by 33 to 22, C by 36 to 19, and D by 28 to 27.

Each of the previous methods has liabilities, and the prior example helps to illustrate what these liabilities are. If we use the most first-place votes to determine the winner, we may end up with someone who is preferred by a small minority and loathed by the majority. If we use the runoff method, it is possible for someone who has almost a clear majority of first-place votes to lose to someone who has an insignificant number of first-place votes. The Survivor method may result in a winner who is clearly preferred to only one of the candidates, who just happens to be the other candidate left when it comes down to a two-person race. The numerical totals method may yield different results depending upon the scoring: a 7-5-3-2-1 scoring system weights first-place votes more heavily than does a 5-4-3-2-1 scoring system. Finally, head-to-head matchups may result in a winner whom the fewest voters feel has the qualities of a leader.

This example has obviously been carefully tweaked: D wins the numerical total by a small amount, and E barely beats D in the head-to-head matchup. Nevertheless, it becomes obvious that in a hotly contested election, the result depends not only on the ballots that were cast, but also on the method for determining the winner. We're back to the question posed at the start of this chapter: Is there a best voting method for determining the outcome of an election?

Imagine for a moment that an election consultant is shown the results of this hypothetical election, and is asked whether he can come up with something better. After studying the example, he might observe that some of the difficulty was engendered by the fact that A is a candidate that polarizes the electorate: eighteen voters prefer A to any other, but the remaining thirty-seven voters rank A dead last. The obvious thing to do would be to devise an algorithm that would help alleviate the problem of polarizing candidates. This can certainly be done, but in doing so our election consultant would undoubtedly notice that no matter what algorithm he devised, other situations that would generally be considered undesirable might arise. In fact, that's exactly what happened to Kenneth Arrow when he started his investigations.

The Impossibility Theorem

Kenneth Arrow was born in New York in 1921. His career was surprisingly similar to Condorcet's: Arrow, like Condorcet, started his career as a mathematician. Like Condorcet, he detoured into economics; and, like Condorcet, it brought him fame and fortune. Like Condorcet, Arrow's work has triggered an intensive investigation of the problem he first brought to attention. One important exception is that as of this writing, Arrow is alive and well, living happily in Palo Alto, and has not yet been forced to flee for his life because he offended the Jacobins or some other political hierarchy. Arrow attended City College in New York, where he majored in mathematics. He continued his mathematical education at Columbia University, obtaining a master's degree, but became interested in economics as the result of meeting Harold Hotelling,[7] a prominent economist and statistician, and decided to obtain a doctorate in economics. World War II intervened, and Arrow served as a weather officer in the Army Air Corps. Arrow spent his tour of duty doing research, eventually publishing a paper on the optimal use of winds in flight planning.

After the war, Arrow resumed his graduate studies, but also worked for the RAND Corporation (one of the first of the think tanks) in Santa Monica, California. He became interested in the problem of constructing methods of translating individual preference rankings into preference rankings for the society. Arrow decided that he would concentrate on those societal ranking methods that were transitive, because transitivity is a property that is easy to express mathematically. This fact readily allows deductions to be made.

Arrow described his progress toward his most famous result, which resembled the efforts of our hypothetical election consultant. At first he tried to devise an algorithm that would eliminate some of the difficulties encountered by existing methods, but after each proposed algorithm eliminated one problem but introduced another, he began to consider the question of whether it was impossible to achieve the desired result.

> I started out with some examples. I had already discovered that these led to some problems. The next thing that was reasonable was to write down a condition that I could outlaw. Then I constructed another example, another method that seemed to meet that problem, and something else didn't seem very right about it. I found I was having difficulty satisfying all of these properties that I thought were desirable, and it occurred to me that they couldn't be satisfied.
>
> After having formulated three or four conditions of this kind, I kept on experimenting. And lo and behold, no matter what I did, there was nothing that would satisfy these axioms. So after a few days of this, I began to get the idea that maybe there was another kind of theorem here, namely, that there was no voting method that would satisfy all the conditions that I regarded as rational and reasonable. It was at this point that I set out to prove it. And it actually turned out to be a matter of only a few days' work.[8]

What were the conditions Arrow had discovered that could not be simultaneously satisfied by any voting method? Arrow's original formulation is somewhat technical;[9] here is a slightly weaker version of Arrow's conditions that is a little more natural than the one that appears in his dissertation.

1. *No voter should have dictatorial powers.* The first condition is something we would certainly want a democracy to have. In other words, when any one individual casts a ballot, the rest of the voters can always vote in such a way that the voting method overrides that individual's preferences.
2. *If every voter prefers candidate A to candidate B, then the voting method must prefer candidate A to candidate B.* The second condition is unanimity. This also seems like an obvious and natural condition for a reasonable voting method to satisfy: if everybody loves it, the society should do it.
3. *The death of a loser should not change the outcome of the election.* At first glance, this condition seems almost unnecessary. We all accept the fact that the death of a winner must necessarily alter the outcome of an election, but how can the death of a loser alter the outcome of an election? To see how this can happen, we'll assume that the death of a

loser simply results in his or her being removed from the ballot. Suppose we have an election with the following collection of ballots.

Number of Ballots	First Choice	Second Choice	Third Choice
40	A	C	B
35	C	B	A
25	B	A	C

This is a very interesting example, for as we shall see, no matter who wins, the death of the "wrong" loser can alter the outcome of the election.

Suppose our voting method results in A winning the election. Now suppose that C dies. Once his name (or letter) is removed from the ballot, we see that $35 + 25 = 60$ people prefer B to A, so B would win.

Suppose instead that our voting method results in B winning, and A dies. In this case, $40 + 35 = 75$ people prefer C to B, so C would win.

Finally, suppose C wins, and B dies. In this scenario, $40 + 25 = 65$ people prefer A to C, again changing the outcome of the election.

No matter what voting method is used to determine the winner, the death of the wrong loser changes the outcome of the election. This is clearly very undesirable.

This condition also brings into focus another aspect of the election process. It is a well-known political aphorism that the presence of a third candidate who has no real chance of winning can have a significant effect on the outcome of an election, as did the presence of Ralph Nader in the 2000 election. This is simply the reverse of the "dead loser" condition above; instead of a losing candidate dying, a candidate who cannot win (and is destined to be a loser) enters an election and changes the outcome. Obviously, we cannot be certain what would have happened in the 2000 election had Nader not been a candidate, but it is generally thought that he drew the great majority of his votes from liberals who would have voted for Al Gore rather than George Bush. Nader received 97,000 votes in Florida, the pivotal state in the election. Bush eventually carried the state by less than 1,000 votes, so Nader's presence probably changed the outcome of the election.

The previous example shows that no voting method can prevent the death of a loser from changing the outcome in an election in which there are three candidates and a hundred voters. But what would happen if there were more candidates, or a different number of voters? What Arrow showed in his famous impossibility theorem, for which he was awarded the Nobel Prize in Economic Science in 1972, was that no transitive voting method can satisfy all three of the above conditions as long as there are at least two voters considering at least three different alternatives.

The Present State of Arrow's Theorem

The noted evolutionary biologist Stephen Jay Gould once proposed a theory he called "punctuated equilibrium,"[10] in which the evolution of species might undergo dramatic changes in a short period of time, and then settle down for an extended period of quiescence. While this theory has yet to be demonstrated to the complete satisfaction of evolutionary biologists, it makes for an accurate description of progress in science and mathematics. This is precisely what happened here. Arrow's theorem, which was undeniably a significant advance, has been followed by a long period during which the result has been extended by relatively small amounts. There have been important developments in related problems, which will be discussed in the next chapter, but Arrow's theorem itself is essentially still the state of the art.

Because Arrow's theorem is a mathematical result, it is interesting to see how mathematicians work with it. The most obvious place to start is to examine the five hypotheses that appear in our formulation of Arrow's theorem. These are: rank ordering of individual and group selections, transitivity of the group selection algorithm, absence of a dictator, unanimity, and the requirement that the death of a loser should not change the outcome of the election.

One of the most intriguing facts about Arrow's theorem is that the five components are independent of each other, but together they are incompatible. One of the first questions a mathematician will ask upon seeing an interesting theorem is, are all the hypotheses required to prove the result? It has been shown that if any one of the components of Arrow's theorem is removed, the remaining four are compatible, in the sense that it is possible to construct voting methods satisfying the four remaining conditions.[11] For instance, if we allow the society to have a dictator (a voter whose ballot is universally adopted), the four remaining conditions are automatically satisfied.

Since any individual's selection is transitive, and the group selection process is simply to adopt the dictator's ballot, the group selection process will be transitive as well. The unanimity requirement will also be satisfied; if everyone agreed that Candidate A was preferable to Candidate B, the dictator also felt this way, and since the dictator's ballot is adopted, the group selection process prefers Candidate A to Candidate B. In practice, near unanimity is not an uncommon occurrence in dictatorships. In pre-war Iraq, Saddam Hussein's choices were reported to have received 99.96 percent of the vote; recently, Bashar al-Assad was reelected president of Syria with 97.62 percent of the vote. Finally, the death of a loser will not affect the outcome of the election, for the removal of the loser from the ballot will not change the dictator's ordering of the remaining candidates, and so the ordering of those other candidates by the society also will not change. So a dictatorship is an example of a "voting method" that satisfies the other four conditions we have been considering.

The unanimity requirement has come in for some mathematical scrutiny that has real-world antecedents. As mentioned previously, in an election with multiple choices, there are frequently options to which a voter is indifferent. One modification of rank ordering, preference intensity, has been discussed previously. At the time of this writing, there are ten announced Republican candidates for president. Senator John McCain, former governor Mitt Romney, and former mayor Rudy Giuliani have attracted the lion's share of the attention. A voter may have decided how to rank these three candidates, but has no strong feelings about any of the other. Some variations of Arrow's theorem replace the condition *if all voters prefer A to B, the voting method prefers A to B* by something along the lines of *if no voter prefers B to A, then the voting method shall not prefer B to A.* This is clearly a modification of the unanimity requirement, to allow for the possibility that a voter may see no reason to prefer A to B, or vice versa, but the modification is not one that most people would regard as a significant change.

One way to dispense with the dilemma presented by transitivity is to simply require that the voting method be able to choose between two alternatives, and not worry about whether the method is transitive or not. Consider once again the situation we encountered with the Condorcet paradox: the majority prefers A to B and B to C, yet prefers C to A. If the voters were ignorant of the results of the A versus B race and the B versus C race, they would not be troubled with the verdict that the majority prefers C to A. Ignorance in this case is bliss, for the subject of the Condorcet paradox never arises. The Condorcet paradox is more likely to trouble social scientists than actual voters, and by simply requiring that the voting method be able to choose between two alternatives, the transitivity problem is eliminated. The head-to head method certainly accomplishes this.

The two components of Arrow's theorem that are most frequently cited as the source of the incompatibility of the five conditions are rank ordering and the dead loser condition. As we have noted, many of the most important elections in the United States simply require the voter to make a single selection, so in practice the Arrow's theorem complications do not arise (although, as will be seen in the next chapter, other complications do). However, the advantages of rank ordering that have been cited (prioritizing and the avoidance of runoff elections) are sufficiently valuable to persuade social scientists (a more pragmatic group than pure mathematicians) to continue to study voting methods using rank ordering.

The dead loser condition is probably the one that most frequently appears in versions of Arrow's theorem—although Arrow himself thought that the dead loser condition was the most pragmatically dispensable of the five components. This raises the question of how we might mathematically assess the value of a voting method—which is itself a subject that is currently being pursued. Just as the first law of thermodynamics compelled us to abandon the quest for free energy from the universe and directed our search toward the maximization of efficiency, Arrow's theorem forces us to search for criteria by which to evaluate voting systems, as there can be no perfect voting system.

The Future of Arrow's Theorem

Niels Bohr's oft-quoted observation, "Prediction is difficult—especially of the future,"[12] is applicable to developments in most scientific endeavors. Even though it is impossible to predict what will happen, there are three possible directions for future results that would raise eyebrows—and possibly win Nobel Prizes if the result were spectacular enough.

As has been observed, most of the results related to Arrow's theorem involve conditions quite similar to Arrow's original ones. Finding an impossibility theorem with a significantly different set of conditions would be extremely interesting, and is a direction which is quite probably being pursued at the moment by social scientists wishing to make a name for themselves. However, just as Arrow found that there was no social preference ranking method satisfying all five conditions, future mathematicians might discover that these conditions, or simple variants of them, might be the only ones that yield an impossibility theorem. The result that it is impossible to find a significantly different impossibility theorem might be even more startling than Arrow's original result.

Finally, one of the ways in which mathematics has always engendered surprises is the variety of environments to which its results are applied. Just as Einstein's theory of relativity proved to be an unexpected and significant application of differential geometry, there may be significant applications of Arrow's theorem (which is, at its core, a result in pure mathematics) in areas vastly different from the social preference setting for which it was originally formulated.

I can't resist the opportunity to insert an idea that has occurred to me (and probably others). One of the consequences of the theory of relativity is that there is no "absolute time"; one observer may see event A as preceding event B, but another observer may see event B as preceding event A. For each observer, the temporal ordering of events is a rank ordering, but the theory of relativity shows that there is no way that the rank ordering of events can be incorporated into a definitive ordering of events on which all observers can agree. Does this sound familiar? It seems a lot like Arrow's theorem to me.

That Reminds Me of a Problem I've Been Working On

Whenever a new result appears in mathematics, especially a breakthrough result such as Arrow's theorem, mathematicians look at it to see whether there is anything about it they can use. Possibly the conclusion of the theorem will supply the vital missing step for a proof, or possibly the proof technique can be adapted to suit their particular needs. A third possibility is a little more indirect: something about the theorem will look familiar. It's not exactly the same problem that is currently stumping them, but there are enough similarities to make them think that with a little tweaking, the theorem is something they can use in one way or another in their own research. And it was a little tweaking of Arrow's theorem that led researchers directly into politics' smoky back rooms.

NOTES

1. See http://www.hoover.org/multimedia/uk/2933921.html.
2. See http://www.cs.unc.edu/~livingst/Banzhaf/.
3. See http://lorrie.cranor.org/pubs/diss/node4.html.
4. See http://www.cooperativeindividualism.org/condorcetbio.html.
5. See http://en.wikipedia.org/wiki/Nicholas_of_Cusa.
6. COMAP, *For All Practical Purposes* (New York: COMAP, 1988). The examples used in this chapter are based on the wonderful textbook *For All Practical Purposes*. If you are going to buy one book from which to continue examining some of the topics in this book, as well as others you might find interesting, I would recommend this one. It was originally constructed by a consortium of teachers of courses with the intention of creating a book that would enable students with a minimal background in mathematics to learn some mathematics that relates to the modern world. It succeeded admirably. Early editions of this book can be purchased on eBay for less than $10.

7. Hotelling is the author of Hotelling's rule in economics, which says that when there is a competitive market for an asset, the asset price will rise at approximately the rate of interest. It sounds good, but my bank pays about 4 percent, and the price at the pump is going up a whole lot faster. A good discussion of this can be found at http://www.env-econ.net/2005/07/oil_prices_hote.html.

8. COMAP, *For All Practical Purposes* (New York: COMAP, 1988; the authors and publishers are the same). This is the book referred to in note 6; there are several editions, and this is the original.

9. K. J. Arrow, "A Difficulty in the Concept of Social Welfare," *Journal of Political Economy* 58(4) (August 1950): pp. 328–46.

10. See http://en.wikipedia.org/wiki/Punctuated_equilibrium.

11. See http://www.csus.edu/indiv/p/pynetf/Arrow_and_Democratic_Practice.pdf.

12. See http://en.wikipedia.org/wiki/Niels_Bohr.

13

The Smoke-Filled Rooms

The Art of the Possible

Otto von Bismarck, the German chancellor, might be most famous for the warrior's approach he took to unifying Germany, but he was a shrewd—and quotable—politician in all respects. "Laws are like sausages: it is better not to see them being made," he once advised. Not surprisingly, then, that he considered politics "the art of the possible."[1] During a long career that saw both military victories and political triumphs (Bismarck was responsible for engineering the unification of Germany), he undoubtedly witnessed and participated in many back-room negotiations. Bismarck probably would have been quite familiar with the following scenario.

A committee to which you belong needs to elect a chair, and you and your fellows have decided to use an instant runoff setup to do so. Four candidates are running for the position. If any candidate receives a majority of the first-place votes, he is elected; otherwise, there is a runoff between the two candidates who received the most first-place votes. You head a faction consisting of four people, and there are two other factions consisting of five and two people, respectively. Of the four candidates for chair, your faction wholeheartedly endorses A, would settle for B, and loathes and detests D.

Everyone in your faction casts an identical ballot: their first choice is A, followed by B, C, and the detestable D. The other two factions have already cast their ballots as follows.

Number of Votes	First Place	Second Place	Third Place	Fourth Place
5	D	C	B	A
2	B	D	A	C

Glumly, you note that if you cast your four ballots, the results will look like this.

Number of Votes	First Place	Second Place	Third Place	Fourth Place
5	D	C	B	A
2	B	D	A	C
4	A	B	C	D

There will be a runoff between A and D, which D will win, 7 to 4. Intolerable. Suddenly, a bright idea occurs to you, and you locate a small room filled with the requisite amount of smoke in which to caucus. You point out to the other members of your faction that if they are willing to switch A and B on their ballots, B will receive six first-place votes (a majority) and win the election. No ballot they cast will enable A to win, but by switching A and B they can ensure an acceptable result and guarantee that D will not win.

This technique also works in more subtle situations. Suppose now that the other seven ballots are tabulated as follows.

Number of Votes	First Place	Second Place	Third Place	Fourth Place
5	D	C	B	A
2	C	B	D	A

If your faction votes as originally intended, the tabulation will look like this.

Number of Votes	First Place	Second Place	Third Place	Fourth Place
5	D	C	B	A
2	C	B	D	A
4	A	B	C	D

In this case, no candidate has a majority, and it now goes to a runoff between A and D as before, which D wins. However, if your faction switches the votes for A and B, the tabulation changes.

Number of Votes	First Place	Second Place	Third Place	Fourth Place
5	D	C	B	A
2	C	B	D	A
4	B	A	C	D

This forces a runoff between B and D, which (mercifully), B wins 6 to 5.

That's sausage making for you. This tactic, which has occurred countless times in the history of elections, is known as insincere voting. Although your faction prefers A, it will settle for B, and the possibility of D winning is sufficiently terrifying that your faction will not vote for its true preferences in order to avoid that outcome.

This example also illustrates two conditions that are necessary in order for insincere voting to be effective: the decision method must be known in advance (in this case, top-two runoff), and the votes of the others must be known in order that a strategy can be accurately plotted. In the example we have been studying, if the votes of the other factions are not known, you could inadvertently undermine your own desires by switching your first-place vote from A to B. The only reason to vote against your preferences is if you know that you can gain by doing so.

Recall that when Kenneth Arrow first began his investigations, he was looking for a system of transferring the preferences of individuals into the preferences of the society, and he attempted to find one that would simultaneously satisfy several apparently desirable attributes. Although Bismarck might have nodded approvingly about how the possible was brought about in the previous example, it is clear that the outcome that was achieved resulted from knowledge of the votes already cast. Intuitively, it seems clear that the knowledge of the votes that have already been cast put those voting last in a more favorable position than those voting first. This certainly seems to add an element of jockeying for position to an election, as well as violating the "one man, one vote" idea that is central to democratic elections: the votes of later voters are worth more than the votes of earlier voters. So the question arises: Does there exist a voting method that eliminates the possibility of insincere voting?

The Gibbard–Satterthwaite Theorem

Like the quest for the perfect system of translating individual preferences into the preferences of the society, the quest for a voting method that eliminates the possibility of insincere voting ends in failure (by this time, you're probably not too surprised that this would be the case). The Gibbard-Satterthwaite theorem[2] states that any voting method must satisfy at least one of three conditions. The phrasing below is slightly different from that in Arrow's theorem, which was stated as "No voting method exists which satisfies . . ."; it is a little easier to phrase the last condition of the Gibbard-Satterthwaite theorem if we state it as "Every voting method must satisfy one of the following conditions." As a result, some of the conditions look like negations of similar conditions in Arrow's theorem.

1. Some voter has dictatorial power. This is the negation of one of the conditions in Arrow's theorem.
2. Some candidate is unelectable. The Gibbard-Satterthwaite theorem does not specify why the candidate is unelectable. It may possibly be that he is extremely unpopular, or is running for an office for which he is not eligible. Or, as has happened in American politics, he may be dead.
3. Some voter with full knowledge of how the other voters will cast their ballots can alter the outcome by switching his or her vote to ensure the election of a different candidate.

The last condition is, of course, the critical one, as it is the essence of insincere voting.

There are two important points to notice about the Gibbard-Satterthwaite theorem. First, it is far more likely that insincere voting will influence the outcome of an election if the number of ballots is relatively small, as it is obviously unlikely that a voter casting a ballot for senator in California (or even Wyoming) can influence the outcome of an election by changing his or her vote. However, there are many elections in which relatively few ballots are cast—chairmanship of committees and nominating conventions are two such examples—and when one considers the possibility of individual voters banding together as a bloc, the scope of the theorem widens significantly.

Additionally, it is highly unlikely that any California voter would have knowledge of how the other voters would cast their ballots; but once again, in small elections such as committee chairmanships or nominating conventions, it is not at all unlikely that there would be voters possessing such knowledge. Voting in the United States Senate also seems subject to insincere voting; there is a time frame established for many votes, and the running tabulation is openly displayed. As a result, even though the Gibbard-Satterthwaite theorem is far less widely known than Arrow's celebrated result, it is nonetheless not only a significant contribution to the social sciences, but also one with considerable real-world ramifications.

Fair Representation

In an ideal democracy, everyone who wishes to participate in the decision-making process would be able to do so by casting a vote. However, the Founding Fathers recognized that most people were too occupied with their lives—with such essentials as farming, shopkeeping, or manufacturing—to be constant participants, and opted for a republic rather than a democracy. In a republic, the voters elect their representatives, who make the decisions.

Unfortunately, under a republican system, it is impossible for all the factions to be fairly represented. A simple example of this might be some political body with five different subgroups, but only three leadership positions. At least two of those subgroups must necessarily be excluded from the leadership.

The American republic has a similar problem. Election to each of our houses of Congress is done on a state-by-state basis. The Senate has two members from each state, so in that house every state has an equal share of leadership positions. Membership in the House of Representatives is a little trickier. The House has a fixed number of representatives (435); the allocation of representatives to each state is done on the basis of the census, which is performed every ten years. It doesn't seem like such a difficult job to determine how many representatives each state should get: if a state has 8 percent of the total population, it should receive 8 percent of the representatives. A quick calculation shows that 8 percent of 435 is 34.8, so the question is whether to round off that .8 of a representative to 34 or 35.

Most elementary school students learn the following algorithm for rounding numbers to whole numbers: round to the nearest integer unless the number to be rounded is midway between two integers (such as

11.5), and then round to the nearest even integer. Using this algorithm, 34.8 is rounded to 35, and 34.5 would be rounded to 34. This is an eminently reasonable algorithm for rounding for the purposes of calculation, but it encounters a problem when rounding for representation in the House of Representatives.

Suppose that the United States consisted of the original thirteen colonies. Twelve of these colonies each have 8 percent of the population; the remaining colony has only 4 percent of the population. According to the above calculation, each of the Big 12 is entitled to 34.8 representatives, which is rounded by the elementary-school algorithm to 35 representatives. The small colony gets only 17.4 representatives, which is rounded down to 17. This procedure designates a total of $12 \times 35 + 17 = 437$ representatives to a House that has room for only 435.

This might seem like a minor problem in number juggling, but the presidential election of 1876 turned on the method by which these numbers were juggled![3] In that year, Rutherford B. Hayes won the election by 185 electoral votes to 184 for his opponent, Samuel Tilden (who, incidentally, won the popular vote by a convincing margin). Had the rounding method used been different, a state that supported Hayes would have received one less electoral vote, and a state that supported Tilden would have received one more. This difference would have swung the election.

The Alabama Paradox[4]

The Founding Fathers recognized the importance of determining the number of electoral votes that each state should receive; indeed, the first presidential veto ever recorded occurred when George Washington vetoed an apportionment method recommended by Alexander Hamilton. (Congress responded by passing a bill to utilize a rounding method proposed by Thomas Jefferson.) Nevertheless, when the disputed election of 1876 occurred, the Hamilton method, also known as the method of largest fractions, was the one being used, having been adopted in 1852.

To illustrate the Hamilton method, we'll start by assuming that we are going to assign representatives to a country that has four states, for a representative body that has thirty-seven members. The following table gives the percentage of the population residing in each of the four states, and also the quota for each state, which is the exact number of representatives to which the state is proportionally entitled.

State	Fraction of Population	Quota (37 × Fraction)
A	.14	5.18
B	.23	8.51
C	.45	16.65
D	.18	6.66

Each quota is a mixed number; an integer plus a fractional part, the fractional part being expressed as a decimal. Each state is initially assigned the integer part of its quota, as indicated by the following table.

State	Quota	Initial Assignment	Remaining Fraction
A	5.18	5	.18
B	8.51	8	.51
C	16.65	16	.65
D	6.66	6	.66

The total initial assignment of representatives is $5+8+16+6=35$, which is 2 representatives short of the desired total of 37. These two representatives are assigned to the states in decreasing order of the leftover fraction. D has the largest leftover fraction (.66), and so gets the first of the two remaining representatives. C has the next largest leftover fraction (.65), and is awarded the second remaining representative. A and B come up short in this process. The final tally is shown in the following table.

State	Percentage of Population	Number of Representatives
A	.14	5
B	.23	8
C	.45	17
D	.18	7

In 1880, C. W. Seaton, the chief clerk of the U.S. Census Office, discovered a curious anomaly in the Hamilton method. He decided to compute the number of representatives each state would receive if the House had anywhere from 275 to 350 seats. In so doing, he discovered that Alabama would have received 8 representatives if the House had 299 representatives, but if the size of the House increased to 300, Alabama would receive only 7 representatives! Thus the Alabama paradox.

299 REPRESENTATIVES

State	1880 Pop	% Pop	Std. Quota	Rounded Quotas	Final Appor	Rank
Kentucky	1648690	3.34	9.99	9	10	0.98
Indiana	1978301	4.01	11.98	11	12	0.98
Wisconsin	1315497	2.66	7.97	7	8	0.97
Pennsylvania	4282891	8.67	25.94	25	26	0.94
Maine	648936	1.31	3.93	3	4	0.93
Michigan	1636937	3.32	9.91	9	10	0.91
Delaware	146608	0.30	0.89	0	1	0.89
Arkansas	802525	1.63	4.86	4	5	0.86
Mississippi	1131597	2.29	6.85	6	7	0.85
New Jersey	1131116	2.29	6.85	6	7	0.85
Iowa	1624615	3.29	9.84	9	10	0.84
Massachusetts	1783085	3.61	10.80	10	11	0.80
New York	5082871	10.30	30.78	30	31	0.78
Connecticut	622700	1.26	3.77	3	4	0.77
West Virginia	618457	1.25	3.75	3	4	0.75
Nebraska	452402	0.92	2.74	2	3	0.74
Minnesota	780773	1.58	4.73	4	5	0.73
Louisiana	939946	1.90	5.69	5	6	0.69
Rhode Island	276531	0.56	1.68	1	2	0.68
Maryland	934943	1.89	5.66	5	6	0.66
Alabama	1262505	2.56	7.65	7	8	0.65

300 REPRESENTATIVES

State	1880 Pop	% Pop	Std. Quota	Rounded Quotas	Final Appor	Difference
Wisconsin	1315497	2.66	7.99	7	8	0.99
Michigan	1636937	3.32	9.95	9	10	0.95
Maine	648936	1.31	3.94	3	4	0.94
Delaware	146608	0.30	0.89	0	1	0.89
New York	5082871	10.30	30.89	30	31	0.89
Mississippi	1131597	2.29	6.88	6	7	0.88
Arkansas	802525	1.63	4.88	4	5	0.88
New Jersey	1131116	2.29	6.87	6	7	0.87
Iowa	1624615	3.29	9.87	9	10	0.87

Massachusetts	1783085	3.61	10.84	10	11	0.84
Connecticut	622700	1.26	3.78	3	4	0.78
West Virginia	618457	1.25	3.76	3	4	0.76
Nebraska	452402	0.92	2.75	2	3	0.75
Minnesota	780773	1.58	4.74	4	5	0.74
Louisiana	939946	1.90	5.71	5	6	0.71
Illinois	3077871	6.23	18.70	18	19	0.70
Maryland	934943	1.89	5.68	5	6	0.68
Rhode Island	276531	0.56	1.68	1	2	0.68
Texas	1591749	3.22	9.67	9	10	0.67
Alabama	1262505	2.56	7.67	7	7	0.67

Those above charts are only half of two, and Seaton had to draw up seventy-five. One can only marvel at the tenacity of Seaton; in those days, in order to crunch the numbers, you *really* had to crunch the numbers without technological assistance, although Seaton may well have had other members of the U.S. Census Office assist him with the computation. Not only that, but one also has to feel sorry for Seaton, who would have been much better served had he been a mathematician or a social scientist. A major discovery like this should be named Seaton's paradox, but that's not what happened. At least he could feel the thrill of discovery.

The paradox can also be seen in this example.

State	House Size 323			House Size 324	
	Fraction of pop.	Quota	Number of reps	Quota	Number of Reps
A	56.7	183.14	183	183.71	184
B	38.5	124.36	124	124.74	125
C	4.2	13.57	14	13.61	13
D	0.6	1.93	2	1.94	2

The Population Paradox

Other defects would appear in the Hamilton method. In 1900, Virginia lost a seat to Maine in the House of Representatives despite the fact that Virginia's population was growing faster than Maine's.

Here's a simple example. Suppose that a state has three districts, the state's representative body has twenty-five members, and the districts are allocated representatives by the Hamilton method.

District	Population (in thousands)	Fraction of State	Quota	Number of Representatives
A	42	10.219	2.55	3
B	81	19.708	4.93	5
C	288	70.073	17.52	17

The next time the census is taken, the population of District A has increased by 1,000, District C has increased by 6,000, while the population of District B is unchanged. The table now becomes

District	Population (in thousands)	Fraction of State	Quota	Number of Representatives
A	43	10.2871	2.57	2
B	81	19.3780	4.84	5
C	294	70.3349	17.60	18

The population of District A has increased by 2.38 percent, whereas the population of District C has increased by 2.08 percent. District A is growing more rapidly than District C, but has actually lost a representative. It would certainly seem fairer that if District C is to gain a representative, it should do so at the expense of District B, which isn't growing at all, and, in fact, could even be shrinking and still receive the same number of seats under the Hamilton method.

The New States Paradox

The Hamilton method failed one last time in 1907, when Oklahoma joined the Union. Prior to Oklahoma's entrance into the Union, the House of Representatives had 386 seats. On a proportion basis, Oklahoma was entitled to 5 seats, so the House was expanded to include $386 + 5 = 391$ representatives. However, when the seats were recalculated, it was discovered that Maine had gained a seat (from 3 to 4), and New York had lost a seat (from 38 to 37).[5]

They're suffering from similar problems in the following example, where a representative house has twenty-nine seats.

District	Population (in thousands)	Quota	Number of Representatives
A	61	3.60	3
B	70	4.13	4
C	265	15.65	16
D	95	5.61	6

Now suppose a new district with a population of 39,000 is added. In a twenty-nine-seat house, its quota is 2.30, so it is entitled to two seats, and a new house is constituted with thirty-one seats. Here is the table that results.

District	Population (in thousands)	Quota	Number of Representatives
A	61	3.57	4
B	70	4.09	4
C	265	15.50	15
D	95	5.56	6
E	39	2.28	2

District A has gained a seat at the expense of District C.

Possibly the Hamilton method could have survived two of the three paradoxes discussed here, but the trifecta killed it. The method currently used, the Huntington-Hill method, adopted in 1941, is a rounding method that is arithmetically somewhat more complex than the Hamilton method. However, as might be suspected, it, too, falls prey to paradoxes. Two mathematical economists, Michel Balinski and H. Peyton Young, were later to show that it couldn't be helped.

The Balinski-Young Theorem

As we have seen, representation is a consequence of the method chosen to round fractions. A quota method is one that rounds the quota to one of the two integers closest to it; for example, if the quota is 18.37, a quota method will round it to either 18 or 19. The Balinski-Young theorem[6] states that it is impossible to devise a quota method for representation that is impervious to both the Alabama paradox and the population paradox.

Although we have introduced this problem in what is probably its most important and controversial context—the structure of the House of Representatives and the Electoral College—the problem discussed here has other important applications. Many situations require quantities to be divided into discrete chunks. As an example, a city's police department has obtained forty new police cars; how should these be assigned to the city's eleven precincts? A philanthropist has left $100,000 in his will to his alma mater for twenty $5,000 scholarships in arts, engineering, and business; how should the twenty scholarships be allocated among these areas? The Balinski-Young theorem shows us that there is no fair way to do this, if we define fairness to mean immunity to the Alabama and populations paradoxes.

Perhaps we should say there is no fair way to do this in the short run. What we mean by this is that there is no way to allocate representatives to states in such a way that every time a census occurs, each state is allocated representatives via a quota method that does not run afoul of the Alabama and population paradoxes. However, there is a method of doing this that will give each state its fair share in the long run. Simply compute the quota for each state and use a randomized rounding procedure to determine whether the number of representatives allocated to that state is the lower or higher of the two possibilities. For instance, if a state has a quota of 14.37 representatives, put 100 balls with numbers 1 through 100 in a jar, blindfold the governor of the state, and have him or her pull out a number. If it is 1 through 37, the state receives 14 representatives; otherwise, it receives 15. In the long run, each state will receive its quota of representatives.

The immediate problem, though, is that this procedure produces Houses of Representatives with varying numbers of representatives. There are fifty states; if each state is awarded the number of representatives equal to the lower of the two possible integers, there would be only 385 representatives. Similarly, there could be as many as 485 representatives. In the long run, of course, there will be an average of 435 representatives.

Recent Developments

Mathematical research is a lot more efficient than when I entered the field in the 1960s. Back then, the institution at which you were teaching subscribed to a number of journals; and most mathematicians had individual subscriptions to the *Notices of the American Mathematical Society*, which printed abstracts of papers published, about to be published or delivered at conferences. If you saw something that interested you, you asked the author for a preprint if the paper wasn't readily available. You read the article, then looked at the bibliography and found other articles of interest, which you xeroxed if they were in your library, or which you obtained by writing the author. Collaboration was still a key aspect of mathematical activity, but it was generally done with colleagues you knew locally or people you had met at conferences.

The Internet completely reshaped the way mathematics is done. The American Mathematical Society maintains MathSciNet,[7] a searchable database of practically every article that has been published in the last fifty years. If you are interested in a particular theorem, such as the Gibbard-Satterthwaite theorem, you simply type it into the MathSciNet search engine—as I just did. Back came a list of sixty-one papers, the most recent being this year, and the earliest in 1975. If the paper was synopsized in *Math Reviews,* it can be obtained and read almost immediately.

This process has made mathematical research much more efficient—and frenetic. The number of publications has jumped exponentially. In addition, the Internet has enabled communication between mathematicians in all parts of the world who might never have come into contact. I have recently collaborated with mathematicians in Germany, Poland, and Greece whom I never would have met (barring a chance meeting at a conference) were it not for the Internet.

MathSciNet also reveals an interesting divergence between the present state of affairs vis-à-vis the Gibbard-Satterthwaite theorem and the Balinski-Young theorem. Insincere voting is related to bluffing in poker; and strategy is a key aspect of game theory, an area of mathematical economics which has resulted in several Nobel Prizes. Of particular interest at the moment is research into areas in which information may or may not be public, such as computing the transactional costs in routing networks. In such a network, the owner of a link is paid for the link's use. A user of the network wants to obtain information, which must be transmitted through a succession of links, at the minimum cost; one obvious strategy is simply to ask the cost of each link. However, the owner of a link may profit by lying about the cost of using his link; this is similar to insincere voting. A key idea being explored is games that are strategy proof; that is, those in which there is no incentive for a player to lie about or hide information from other players. It is easy to see how this is related to voting.

On the other hand, only twenty-two articles are listed on MathSciNet concerning the Balinski-Young theorem, the most recent being in 1990. The field is evidently dormant, despite the fact that there is an obvious gap in the area; I have not been able to find any work incorporating the new states paradox in Balinski-Young type theorems. Nonetheless, because of the importance of the Electoral College, the current (Huntington-Hill) method of selecting representatives is currently being investigated[8] by mathematicians and political scientists to see if better methods are available.

As is often the case, when an ideal result is shown to be impossible, it is important to develop criteria for the evaluation of what can be achieved under various circumstances. An impossible result establishes budgetary constraints, and it is up to us to determine what to optimize, and how to accomplish that, while living within our budget.

NOTES

1. See http://www.brainyquote.com/quotes/authors/o/otto_von_bismarck.html. I'm a big fan of quotes, and this site has a lot of great ones.
2. Allan Gibbard is a professor of philosophy at the University of Michigan, and Mark Satterthwaite is a professor of strategic management and managerial economics at Northwestern University. Despite the Midwestern locales of these two universities, the Gibbard-Satterthwaite theorem was not hatched over a dinner table while the two were discussing insincere voting. The original result is due to Gibbard; the improvement to Satterthwaite, as the following papers indicate: Allan Gibbard, "Manipulation of Voting Schemes: A General Result," *Econometrica* 41 (4) (1973): pp. 587–601; Mark A. Satterthwaite, "Strategy-proofness and Arrow's Conditions: Existence and Correspondence Theorems for Voting Procedures and Social Welfare Functions," *Journal of Economic Theory* 10 (April 1975): pp. 187–217.
3. See http://en.wikipedia.org/wiki/United_States_presidential_election,_1876. An interesting sidelight to the election is that there was a minor third party in this election called the Greenback Party. Insert cynical remark here.
4. See http://occawlonline.pearsoned.com/bookbind/pubbooks/pirnot_awl/chapter1/ custom3/deluxe-content.html#excel. This site has Excel spreadsheets you can download for both the Alabama paradox and the Huntington-Hill apportionment method.
5. See http://www.cut-the-knot.org/ctk/Democracy.shtml. This site not only has explanations of all the paradoxes, but nice Java applets that you can use to see them in action.
6. M. L. Balinski and H. P. Young, *Fair Representation,* 2nd ed. (Washington, D. C.: Brookings Institution, 2001). Unlike the authors of the Gibbard-Satterthwaite theorem, who were separated by time and probably distance, Balinski and Young were together at New York University for much of the period during which the relevant ideas were formulated and the Balinski-Young theorem proved.
7. MathSciNet is a wonderful database, but you either have to belong to an institution that subscribes to it (many colleges and universities, as well as some research-oriented businesses, are subscribers), or have a tidy chunk of change burning a hole in your pocket.
8. See http://rangevoting.org/Apportion.html.

14
Through a Glass Darkly

The Half-Full Glass

Although mathematics and physics have shown us that there are things we cannot know and feats we cannot accomplish, just because utopia is unattainable does not mean that dystopia is inevitable. I was born in a world just coming into the electronic age, at a time when the values cherished by the Western democracies were threatened as never before and never since, so when I look at what the world today has to offer and the threats it presents, it seems to me that the glass is much more than half full.

The sciences, along with their common language of mathematics, will continue to investigate the world we know and the worlds we hypothesize. Along with future discoveries will come future dead ends, which will also serve to tell us more about the universe. I feel that this book would be incomplete without some sort of summary of what we have learned about the limitations of knowledge, but I also feel that it would be incomplete if it did not make an attempt to foresee what the future may hold in this area. An added reason for doing so is that it is unlikely that I will be remembered for any sort of spectacular success in my later years; maybe I can be remembered for a spectacular failure, like Comte or Newcomb. After all, we live in a society in which notoriety and fame are often confused.

Besides, mathematics is an area in which a really good question can achieve so much publicity that its eventual resolution, and those who resolve it, almost become historical footnotes to the question itself. Pierre de Fermat, Bernhard Riemann, and Henri Poincaré are among the greats of mathematics—but Fermat is almost certainly best remembered for Fermat's last theorem, Riemann for the Riemann hypothesis, and Poincaré for the Poincaré conjecture. Fermat's last theorem fell a decade ago to Andrew Wiles, who was denied a Fields Medal for his accomplishment because he was too old (Fields Medals are reserved for thirtysomethings and twentysomethings, and Wiles missed it by a year or so). The Poincaré conjecture succumbed more recently, and argument still exists in the mathematical community as to who should get the lion's share of the credit, with the Russian mathematician Grigori Pereleman in the lead. The Riemann hypothesis is still just that—a hypothesis. Besides, one can go into the annals with a great conjecture, even if one is not a great mathematician. Most mathematicians would be hard-pressed to name a single one of Christian Goldbach's mathematical accomplishments,[1] but everyone knows Goldbach's conjecture—the elegantly simple "every even number is the sum of two primes"—a conjecture understandable to grade-school children but still standing unproved after more than a quarter of a millennium of effort.

The Impact of Age

There is a perception that mathematicians and physicists do their best work before they are thirty. That's not necessarily true, but it is true that the young make a disproportionate contribution to these subjects. This may be because the young are less willing to accept the generally accepted paradigms. Unquestionably, age confers both disadvantages and advantages.

Sometimes these disadvantages force practitioners into other areas. It is said with some truth that as physicists age, they become philosophers. They tend to pay less attention to discovering the phenomena of reality than to reflecting on the nature of reality. With the possibility of multiple dimensions and the nature of quantum reality still unresolved, there is no question that there is considerable room for reflection.

When I was young, like most boys growing up in my era, I was extremely interested in both sports and games—the difference being that in a sport you keep score and sweat from your own exertion, and in a game you merely keep score (sorry, Tiger, golf is a game, not a sport). One of my interests was chess; I studied the game avidly and read stories about it. I recall one story about a chess grand master traveling incognito on a train, and who is drawn into a pickup game with an up-and-coming young phenom. At one stage, the grand master observes that as one ages, one no longer moves the pieces, one watches them move.[2]

I think there is a profound amount of both truth and generality in that remark; I believe it applies to mathematicians as well as grand masters, and I certainly feel that it applies to me. I am no longer capable of constructing lengthy and complicated proofs, but I have acquired a good deal of "feel" for what the right result should be. Some years ago, I had the pleasure of working with Alekos Arvanitakis, a brilliant young mathematician from Greece. I have never met Alekos; he contacted me because he had obtained results relevant to a paper that I had written, and we started to e-mail and eventually began collaborating. He brought new insight and considerable talent to a field I felt I had been worked out (in the sense that nothing really interesting was left to prove). I would suggest something that felt like it ought to be true, and within a week Alekos had e-mailed me a proof. I felt somewhat guilty about coauthoring the paper, feeling that Alekos had done most of the heavy lifting, but I decided that at least I had a sense of what needed to be lifted. I could no longer move the pieces as well as I did when I was younger, but I could watch them move as if of their own volition.

Classifying the Dead Ends

Looking back through the previous chapters, it seems to me that the problems and phenomena we have investigated fall into a number of distinct categories.

Of these, the most ancient are the problems that we are unable to solve within a particular framework. The classic examples of these are problems such as the duplication of the cube and the roots of the quintic. In both cases, the problem was not so much an inability to solve the problem as it was to solve the problem using given tools. The usual way to deal with such problems is to invent new tools. That's exactly how the cube was duplicated and the roots of the quintic found: by using tools other than those available to formal Euclidean geometry, and finding ways other than solutions by radicals to express certain numbers.

Undecidable propositions could well belong in this category. Recall that

Goodstein's theorem was undecidable in the framework provided by the Peano axioms, but admitted a solution when the axiom of infinity from Zermelo-Fraenkel set theory was incorporated into the axiom set. This raises an obvious question: Is the nature of undecidability simply a matter of choosing the correct axioms, or the correct tools?

Or are there genuinely undecidable propositions that do not lie within the reach of any consistent axiom set?

A second category of insoluble problems exists because of the inability to obtain adequate information to solve the problem. Failure to obtain this information may occur because the information simply does not exist (many quantum-mechanical phenomena come under this heading), because it is impossible to obtain accurate enough information (this describes random and chaotic phenomena), or because we are exposed to information overload and simply cannot analyze the information efficiently (this describes intractable problems).

We come now to the third category of problems we cannot solve: those in which we are asking for too much. So far, the most significant problems we have found in this area are the ones from the social sciences, involving the quest for voting systems or systems of representation. There are innumerable formal problems that fall under this description, such as the problem of covering the chessboard with the two diagonal squares removed with 1×2 tiles, and possibly the techniques involved for analyzing such problems will be useful in more practical situations.

Finally, there are the questions that turn out to have several right answers. The independence of the continuum hypothesis and the resolution of the dilemma posed by the parallel postulate fall into this category. It seems reasonably safe to predict that there will be other surprises awaiting us; questions to which the answers lie outside the realm of what we would expect, including the possibility that there are questions whose answers depend upon the perspective of the questioner. For example, the theory of relativity answers the riddle of which came first, the chicken or the egg, with the answer that it depends upon who's asking the question—and how fast and in what direction they are moving.

There is one last recourse when we are absolutely, completely, and totally stymied: try to find an approximate solution. After all, we don't need to know the value of π to a gazillion decimal places; four decimal places suffice for most problems. Although it is impossible to find exact solutions by radicals to certain quintics, one can find rational solutions to any desired degree of accuracy. It is extremely important to be able to do this. Salesmen, after all, are very likely to continue traveling even in the absence of a polynomial-time solution to the traveling salesman problem, and if we can find a polynomial-time algorithm approximating the exact solution within a few percent, we can save substantially on both fuel and time. "Good enough" is sometimes more than good enough.

Two Predictions I Feel Are Likely

A common theme that emerges from many of the problems that we have examined is that if the system we are describing is sufficiently complex, there will be truths that we will be unable to ascertain. Of course, the paramount example of this is Gödel's result on undecidable propositions, but I would expect that mathematicians and logicians of the future would be able to do one of two things: either describe what makes an axiomatic system sufficiently complex to admit undecidable propositions, or show that such a description is impossible. There may already be partial results in this area, but if the latter result had been proven, I think it would be a sufficiently breathtaking result that it would be widely known in the mathematical community. I think the same will hold for Hilbert's quest to axiomatize physics: either it will be shown that this cannot be done successfully, or if successful, it will result in physical analogues of undecidable propositions. The existence of analogues of the uncertainty principle will arise not from a quantum hypothesis (although that hypothesis specifically resulted in the uncertainty principle), but from the axiomatization itself. That axiomatization will show that there must be results along the lines of the uncertainty principle, but it will not tell us what those results are.

I admit there is a certain vagueness about the predictions made in the previous paragraph, so I will offer a more concrete one. It will be shown for every one of the myriad of NP-hard problems that any polynomial-time algorithm devised for solving it admits anomalies of the form we encountered when discussing how priority-list scheduling can lead to situations in which making everything better makes things worse. This was also seen when we applied the nearest neighbor algorithm to the traveling salesman problem; it is easy to construct an array of towns and distances between them such that, if all the distances were shortened, the nearest neighbor algorithm resulted in a path with a longer total distance than the total distance given by that algorithm for the original configuration.

If I were a young, but tenured, specialist in this area (the specification of tenure is necessary because this might be a problem requiring an immense amount of time, and you don't want to risk your chance of tenure on a problem for which you might not achieve quick results), or a mature specialist who was looking for an eye-opening result, I'd give this one a shot. After all, it doesn't seem so unlikely that Cook's techniques for demonstrating the equivalence of NP-hard problems could be modified to show that a hole in one algorithm must necessarily result in holes in others. I'm not capable of moving these pieces, but I really believe that I can see them move.

Falling Off the Train

Whenever one considers memorable prognostications that have proven to be incredibly wrong, one has to at least mention a classic that occurred a couple of decades ago. The Soviet Union had just collapsed, the United States was the world's only superpower, and Francis Fukuyama produced a widely publicized essay entitled "The End of History?" This less-than-prescient comment is taken from that essay: "What we may be witnessing is not just the end of the Cold War, or the passing of a particular period of post-war history, but the end of history as such: that is, the end point of mankind's ideological evolution and the universalization of Western liberal democracy as the final form of human government."[3]

Karl Marx may have been discredited as an economic theorist, but he really nailed it this time with his observation that when the train of history rounds a corner, the thinkers fall off.[4] Even conservatives would likely welcome the universalization of Western liberal democracy as the final form of human government, but the events of the last two decades have shown that the millennium, at least in the sense of the ultimate fulfillment of Fukuyama's prediction, is not yet at hand.

In retrospect, Isaac Asimov had a much clearer view of how history unfolds. Asimov may not have been the first of the great popularizers of science (my nominee would be Paul de Kruif, author of the classic *Microbe Hunters*), but he was undoubtedly the most prolific. He has works in every major category of the Dewey decimal system except philosophy, which he perhaps resisted because his academic background was in biochemistry rather than physics. It is rather surprising that his popularized science works are written as rather straightforward presentations of facts ("the moon is only one-forty-ninth the size of the Earth, and it is the nearest celestial body"), because his fame originally came from his highly entertaining and often-prescient science fiction. He was one of the three great early writers of science fiction (Arthur C. Clarke and Robert Heinlein being the others), and his ideas were often unbelievably ingenious. One of his earliest published stories, "Nightfall," describes the difficulty experienced by a civilization in trying to discover the mysteries of gravitation in a planetary system with six nearby stars. I found particularly amusing a story he wrote in which he described the logical consequences surrounding the discovery of thiotimoline, a substance that dissolved in water 1.2 seconds *before* the water was added.

Asimov's best-known major work of science fiction is the Foundation trilogy.[5] In it, a future interstellar empire is collapsing, a collapse foreseen by Hari Selden, a mathematical sociologist who uses statistics to predict the future history of the empire. Selden's computations are upset by the arrival of the Mule, a mutant whose special psychic abilities enable him to seize and hold power. Others, such as Georg Hegel, had emphasized that history is shaped by specific individuals, whom he called "world-historical individuals." Asimov's contribution was to note that the propensity of civilizations to produce such individuals may invalidate any hope of applying techniques, such as those used in statistical mechanics, to history. After all, no single air molecule has the capability to alter the behavior of large quantities of other air molecules the way a single individual can alter the course of history.

Does this mean it is impossible to come up with a mathematical scheme for assessing or predicting history? It's an interesting question, and I am unaware of any Hari Seldens, past or present, who have attempted any sort of scheme with any success. Possibly, when the mathematics of chaos is sufficiently well developed, it may be possible to set some sort of limits on what might be accomplished in this area.

In the 1960s, the French mathematician René Thom inaugurated a branch of mathematics now called catastrophe theory.[6] This was an attempt to analyze dramatic changes in behavior of phenomena arising from small changes in the parameters describing the phenomena. Sound familiar? It certainly has a good deal of the flavor of chaos theory. Additionally, catastrophe theory looks at nonlinear phenomena, as does much of chaos theory. A major difference, however, is that catastrophe theory views dramatic shifts in behavior of the underlying parameters as manifestations of standardized geometrical behavior in a larger parameter space. From a practical standpoint in actually predicting impending catastrophes, this isn't much help. It might be nice to know that the next stock market collapse is simply the expected behavior of a well-defined geometrical structure in a higher-dimensional space, but unless we can get a handle on exactly what the parameters governing that higher-dimensional space measure, and do so in some a priori manner, it's not very useful.

There are, of course, numerous applications of mathematics in the social sciences; courses in these applications are offered at almost every institution above the secondary level. Some have proved remarkably successful, primarily in areas offering easy quantification of the relevant parameters. However, just because it is easy to quantify the relevant parameters does not guarantee success; almost all the major stock market crashes have been characterized by an utter inability on the part of the major prognosticators to predict them. Possibly, some future Hari Selden may glimpse the multidimensional geometrical structures in whose shapes are written the portents of the future, but I think it more likely that some future Kenneth Arrow may discover that even though those geometrical structures may exist, there is no way for us to determine what they are.

In the Footsteps of Aquinas

Some of the greatest minds in history have endeavored mightily, as did Saint Thomas Aquinas, to prove the existence of a deity; and some equally great minds have endeavored just as mightily to prove that a deity cannot exist. These proofs have one thing in common. They have utterly failed to convince the other side.

It is hard to imagine a proof on any subject that would elicit greater interest on the part of the public. Such a proof would answer, one way or another, one of the most profound questions that has ever been asked. It is also likely that the appearance of such a proof would generate a firestorm of controversy as to its validity. It is unlikely that such a proof would be a simple one, as most of the simple lines of proof have been exhausted centuries ago.

I've seen several of these proofs employing dubious hypotheses and/or dubious logic, although I have yet to see any proofs on either side employing purely mathematical reasoning, with numbers, shapes, tables, or any of the other concepts of mathematics. Possibly the easiest is the one that argues for the nonexistence of God via the following paradoxical construction: If God exists, he or she must be all-powerful, so can God make a stone so heavy that he or she cannot lift it? If he or she cannot make such a stone, then he or she cannot be all-powerful. If he or she can make such a stone, then the fact that he or she cannot lift it provides evidence that he or she cannot be all-powerful.

There is a limit even to the all-powerful, and overcoming such a paradox is one of them. One might with equal validity argue that the inability to duplicate the cube using compass and straightedge proves the nonexistence of God.

In fairness, a rebuttal should be given for one of the classic arguments given for the existence of God; the "first cause" argument. It is argued that something cannot arise from nothing, and therefore something must have been here first, that something being God. It sounds convincing, but it simply doesn't hold up. One current cosmological theory postulates the existence of an eternal multiverse. Our universe arose from the big bang some 13 billion years or so ago, but these events may have occurred infinitely often previously in a multiverse that has existed forever. At present, there is simply no way to know.

Both sides have been so busy trying to construct proofs supporting their case that it seems to me they have overlooked the obvious. Once the attributes of a deity are precisely defined, there may be a proof that it is impossible to prove the existence or nonexistence of such a deity. Alternatively, the deity hypothesis may possibly be shown to be independent of a set of philosophical axioms, adjoining either the deity hypothesis or its negation to those axioms leads to a consistent axiom set.

I must admit to a bias in favor of such a resolution. An awful lot of intellectual firepower has been brought to bear on this issue, but as yet no direct hits have been scored. I think society would be better served if individuals with the ability to make headway on such a problem devoted themselves to finding cures for AIDS or bird flu. This is probably a pipe dream on my part, a delusion somewhat substantiated by the fact that even though it has been known for centuries that it is impossible to trisect the angle using compass and straightedge, probably thousands of individuals are even now struggling to achieve the impossible. I shudder to think how many people might devote themselves to attempting to disprove a result such as the independence of the deity hypothesis.

I Know What I Like

We decorate our residences and offices with pictures, and we surround ourselves with music. Despite the obvious and nearly universal appeal of the visual and auditory arts, I stand (or sit) with Rex Stout's corpulent detective Nero Wolfe, who once stated that cooking is the subtlest and kindliest of the arts. For me, the ethereal beauty of Monet's water lilies, or the transcendent majesty of a Beethoven symphony, pales in comparison to a steaming bowl of hot and sour soup, followed shortly thereafter by a succulent dish of kung pao chicken (extra spicy).

Much though I love Monet, Beethoven, Nero Wolfe, and Chinese cuisine, these passions are not universally shared. In fact, Arrow's theorem sheds some light on artistic (and culinary) preferences of a group; there is no way to translate individual preferences in these areas into a societal ranking consistent with the five conditions set down in Arrow's theorem.

However, just as political success awaits the candidate who can appeal to the majority, fame and fortune undoubtedly await the individual who discovers the key to creating widely appreciated art, music—or food. Mathematics has met with little success in this area.

Garrett Birkhoff was one of the preeminent American mathematicians of the first half of the twentieth century. He made noteworthy contributions to celestial mechanics, statistical mechanics, and quantum mechanics, in addition to his work in pure mathematics. Generations of college students—including mine—learned the theories of groups, rings, and fields from his landmark text on abstract algebra,[7] coauthored with the equally eminent Saunders Mac Lane.

Birkhoff also had a keen interest in aesthetics, and attempted to apply mathematics to the evaluation of art, music, and poetry. To be fair, his efforts were nowhere near as laughable as the reaction of Charles Babbage, one of the pioneers in the construction of mechanical computational devices. On reading a poem by Tennyson that included the line "Every moment dies a man,/Every moment one is born," Babbage sent a note to Tennyson pointing out that, to be strictly accurate, Tennyson should have written, "Every moment dies a man,/Every moment one and one-sixteenth is born."

Birkhoff's basic formula for computing aesthetic value was that the aesthetic measure of a work of art was equal to the quotient of its aesthetic order divided by its complexity—orderly things were beautiful, complex things were not. The mathematicians whose musical tastes I have ascertained generally seem to conform to this rule; Bach generally receives a better reception among mathematicians than does Shostakovitch. In fact, when a friend introduced me to a Bach chaconne, he started by describing it by saying that it has 256 measures ($256 = 2^8$) divided into 4 sections of 64 measures ($64 = 2^6$), and I liked it even before I heard a single note.

To some extent, the idea that order is more attractive accords with statistical surveys that have determined some fairly obvious broad generalities in aesthetics: the majority prefers symmetry to asymmetry, pattern to absence of pattern. However, some of Birkhoff's subsidiary formulas are almost painful to read. For example, to compute the aesthetic order of a poem, Birkhoff devised the formula $O = aa + 2r + 2m - 2ae - 2ce$, where aa stands for alliteration and assonance, r for rhyme, m for musical sounds, ae for alliterative excess, and ce for excess of consonant sounds. To be fair to Birkhoff, his efforts antedate Arrow's theorem by decades, and he admitted that intuitive appreciation outweighed mathematical calculation. Nonetheless,

Birkhoff believed that the intuitive appreciation stemmed from an unconscious application of the mathematical aspects of his formulas.

If I were to hazard a guess, the complexity of aesthetic factors is highly likely to serve as a barrier to any sort of aesthetic predictability. Evidence for this is the total inability of husbands to predict what their wives will appreciate; as against that, wives often seem to have an uncanny ability to know what their husbands will like. If there's a theorem in here somewhere, it wouldn't surprise me in the least if it is a woman who finds it.

The Ultimate Questions

Is it possible for mathematics to come up with a way to know where the dead ends are, or what we cannot know? This book is filled with specific examples of dead ends that we have circumvented and things we have found that we cannot know, but is it conceivable that there exists a meta-theorem somewhere that delineates some of the characteristics of mathematical or scientific ideas that are beyond the reach of knowledge? Or is there a meta-theorem that says it is impossible for a meta-theorem as described in the previous sentence to exist?

I think that results in this area are unlikely to be so grandiose, and that the dead ends and limits to knowledge will arise only in specific circumstances rather than as the result of an ultimate meta-theorem about the limits of knowledge. Mathematics can only discuss mathematical objects; although the scope of what constitutes mathematical objects is continually expanding. As great a mathematician as Gauss was, he did not foresee the possibility of treating infinities as completed quantities, and infinities are clearly something he could have considered to have the potential for being mathematical objects. We do not yet have the mathematical objects needed to discuss art, or beauty, or love; but that does not mean they do not exist, only that if they exist, we haven't found them. Indeed, if we exist in Tegmark's Level 4 multiverse, which consists of realizations of mathematical objects, then since art, beauty, and love exist, they are mathematical objects; we just have not found the way to describe them mathematically. Maybe Keats really was right about beauty being truth, and vice versa; most mathematicians believe at least half of it, that truth is beauty. If some future Kurt Gödel manages to construct a mathematical theory of interpersonal relationships, and in so doing proves that there are aspects to love that we cannot know, how deliciously ironic it would be that mathematics could prove what poets, philosophers, and psychologists have only been able to conjecture.

NOTES

1. See http://www-history.mcs.st-andrews.ac.uk/Biographies/Goldbach.html. I couldn't resist and looked up Goldbach's biography. He knew a lot of the greats and actually did some useful mathematics, but never in the article did I see the word *dilettante*, which, it seemed to me, best described him.
2. Although I couldn't locate a copy of the book, I recall the story being in I. Chernev and F. Reinfeld *The Fireside Book of Chess* (New York: Simon & Schuster, 1966).
3. See http://www.wesjones.com/eoh.htm.
4. See http://www.facstaff.bucknell.edu/gschnedr/marxweb.htm.
5. I. Asimov *Foundation* (New York: Gnome Press, 1951); *Foundation and Empire* (New York: Gnome Press, 1952); *Second Foundation* (New York: Gnome Press, 1953). Also see http://www.asimovonline.com/asimov_home_page.html. This is the home page for a complete introduction to Isaac Asimov. One could spend the better part of a lifetime reading his books and short stories, and it would probably be the better part of the reader's lifetime.
6. See http://en.wikipedia.org/wiki/Catastrophe_theory. This provided an introduction to catastrophe theory, along with a description of various types of catastrophes. Regrettably, there are no predictions of future catastrophes.
7. G. Birkhoff and S. Mac Lane, *Algebra* (New York: Macmillan, 1979). This is a later edition of the book that I used.

Index

Anderson, Carl, 135–36

Annals of Mathematics, 32

antielectron, *see* positron

Antoinette, Marie, 81

Apple II computer, 126

approximate solutions, 82, 149–50, 168, 178, 179, 240–41

Aquinas, Saint Thomas, 244

Archimedes, xiii, xiv, 71, 72

Archytas, 69, 75

Arens, Richard, 43

arithmetic, 68

    consistency of arithmetic axioms, 119–22, 123–24

    fundamental theorem of, 42

    transfinite, 19

Arrow, Kenneth, xv, xvii, 214–15, 225

Arrow's theorem, xv–xvi, xvii, 9, 214–20, 245

    conditions of, 215–16, 217–19

    future of, 219–20

    present state of, 217–19

Ars Magna (Cardano), 88

Arvanitakis, Alekos, 239

ASCII, 174

Asimov, Isaac, 242–43

Aspect, Alain, 62

associative law, 90, 91, 92

astigmatism, 47

astrology, 87

astronomy, 30

Athens, plague in, 68–69

atomic bomb, 142

atomic theory, 134

Australia, rank-order voting in, 209, 211

automobile repairs, scheduling of, 3–7

axiom of choice, 21–23, 24

B

Babbage, Charles, 246

Babylonians, solution of quadratic equations by, 83

back-room voting, *see* insincere voting

Bacon, Francis, 19

Bade, William, x, 122

baker's transformation, 177–78, 178

Balinski, Michel, 233

Balinski-Young theorem, 233–34, 235

Banach algebras, 37

Banach-Tarski paradox, 23

Banzhaf Power Index (BPI), 206–7

base-2 number system, *see* binary (base-2) number system

base-10 number system, *see* decimal (base-10) number system

basic objects, 102

beam splitters, 50–51

Beane, Billy, 8

Beethoven, Ludwig von, 85, 245

Bell, John, 61–62

Bell's theorem, 61–62

Beltrami, Eugenio, 112

Bernoulli, Johann, 106

Berra, Yogi, 158

Bessel, William, 97

Bessel functions, 97

beta decay, 148

big bang theory, 36, 38, 245

bill paying, 158

binary (base-2) number system, 172–73

    normal number in base 2, 173, 175, 176

Birch and Swinnerton-Dyer conjecture, 1–2

Birkhoff, Garrett, 246

Bismarck, Otto von, 223, 225

blackbody, 44

black holes, 143, 196–97